Cross-Platform Development with Qt 6 and Modern C++

Design and build applications with modern graphical user interfaces without worrying about platform dependency

Nibedit Dey

BIRMINGHAM—MUMBAI

Cross-Platform Development with Qt 6 and Modern C++

Group Product Manager: Aaron Lazar
Publishing Product Manager: Kushal Dave
Senior Editor: Storm Mann
Content Development Editor: Nithya Sadanandan
Technical Editor: Gaurav Gala
Copy Editor: Safis Editing
Project Coordinator: Deeksha Thakkar
Proofreader: Safis Editing
Indexer: Vinayak Purushotham
Production Designer: Aparna Bhagat

First published: May 2021

Production reference: 2230721

Published by Packt Publishing Ltd.
Livery Place
35 Livery Street
Birmingham
B3 2PB, UK.

ISBN 978-1-80020-458-4

www.packt.com

To my mother for believing in my dreams and to my father for his sacrifices and supporting me through my ups and downs.

To my childhood English teacher Mr. Harendra Das, who laid the foundation and honed my English communication skills, and my Science teacher Mr. Jayanta Kumar Das, who introduced me to computer and used to say "You will write a book one day"!

– Nibedit Dey

Contributors

About the author

Nibedit Dey is a software engineer turned serial entrepreneur with over a decade of experience in building complex software-based products with amazing user interfaces. Before starting his entrepreneurial journey, he worked for Larsen & Toubro and Tektronix in different R&D roles. He holds a bachelor's degree in biomedical engineering and a master's degree in digital design and embedded systems. Specializing in Qt and embedded technologies, his current role involves end-to-end ownership of products right from architecture to delivery. Currently, he manages two technology-driven product start-ups named ibrum technologies and AIDIA Health. He is a tech-savvy developer who is passionate about embracing new technologies.

About the reviewers

Andrey Gavrilin is a senior software engineer in an international company that provides treasury management cloud solutions. He has an MSc degree in engineering (industrial automation) and has worked in different areas such as accounting and staffing, road data banks, web and Linux distribution development, and fintech. His interests include mathematics, electronics, embedded systems, full-stack web development, retro gaming, and retro programming.

Syed Aftab has a bachelor's degree in engineering with a focus on electronics and communications. He has around 16 years of experience in software product development. Syed is skilled in software development using C and C++ technologies on various platforms such as Unix, Linux, Windows, and embedded platforms.

Programming, sharing his programming experience, and mentoring juniors are his passions. You can get in touch with him at https://www.linkedin.com/in/syed-aftab-a06a1943/

Table of Contents

3

GUI Design Using Qt Widgets

4

Qt Quick and QML

Section 2: Cross-Platform Development

5

Cross-Platform Development

Section 3: Advanced Programming, Debugging, and Deployment

6

Signals and Slots

9
Testing and Debugging

10
Deploying Qt Applications

11
Internationalization

12
Performance Considerations

Other Books You May Enjoy

Index

Preface

Qt is a cross-platform application development framework designed to create great software applications with amazing user interfaces for desktop, embedded, and mobile platforms. It provides developers with a great set of tools for designing and building great applications without having to worry about platform dependency.

In this book, we will focus on Qt 6, the latest version of the Qt framework. This book will help you with creating user-friendly and functional graphical user interfaces. You will also gain an advantage over competitors by providing better-looking applications with a consistent look and feel across different platforms.

Developers who want to build a cross-platform application with an interactive GUI will be able to put their knowledge to work with this practical guide. The book provides a hands-on approach to implementing the concepts and associated mechanism that will have your application up-and-running in no time. You will also be provided explanation for essential concepts with examples to give you a complete learning experience.

You will begin by exploring the Qt framework across different platforms. You will learn how to configure Qt on different platforms, understand different Qt modules, learn core concepts, and learn how they can be used to build efficient GUI applications. You will be able to build, run, test, and deploy applications across different platforms. You will also learn to customize the look and feel of the application and develop a translation aware application. Apart from learning the complete application process, the book will also help you in identifying the bottlenecks and how to address them in order to enhance the performance of your application.

By the end of this book, you will be able to build and deploy your own Qt applications on different platforms.

Who this book is for

This book is intended for developers and programmers who want to build GUI-based applications. It is also intended for software engineers who have coded in C++ before. The entry barrier isn't that high, so if you understand basics C++ and OOPS concepts, then you can embark on this journey.

In addition, this book can help intermediate-level Qt developers, who want to build and deploy in other platforms. Working professionals or students, who want to get started with Qt programming, as well as programmers who are new to Qt, will find this book useful.

What this book covers

Chapter 1, Introduction to Qt 6, will introduce you to Qt and describe how to set it up on a machine. By the end of the chapter, readers will be able to build Qt from source code and get started on their platform of choice.

Chapter 2, Introduction to Qt Creator, introduces you to the Qt Creator IDE and its user interface. This chapter will also teach you how to create and manage projects in Qt Creator. You will learn to develop a simple *Hello World* application using Qt Creator and learn about different shortcuts, and practical tips.

Chapter 3, GUI Design Using Qt Widgets, explores the Qt Widgets module. Here, you will learn the various kinds of widgets that are available for creating GUIs. You will also be introduced to Layouts, Qt Designer, and learn how to create your own custom controls. This chapter will help you in developing your first GUI application using Qt.

Chapter 4, Qt Quick and QML, covers fundamentals of Qt Quick and QML, Qt Quick Controls, Qt Quick Designer, Qt Quick Layouts, and Basic QML Scripting. In this chapter, you will learn to use Qt Quick controls and how to integrate C++ code with QML. By the end of this chapter, you will be able to create a modern application with fluid UI using QML.

Chapter 5, Cross-Platform Development, explores cross-platform development using Qt. You will learn about different settings in Qt Creator. In this chapter, you will be able to run sample applications on your favorite desktop and mobile platforms.

Chapter 6, Signals and Slots, covers the signals and slots mechanism in depth. You will be able to communicate between different C++ classes and between C++ and QML. You will also learn about events, event filters and event loop.

Chapter 7, Model View Programming, introduces you to the Model/View architecture in Qt and its core concepts. Here, you will be able to write custom models and delegates . You can use these to display required information on your Qt Widget-based or Qt Quick-based GUI application.

Chapter 8, Graphics and Animations, introduces the concepts of 2D graphics and animations. You will learn how to use painter APIs to draw different shapes on the screen. We will further discuss the possibility of graphics data representation using Qt's Graphics View framework and Scene Graph. This chapter will guide you towards creating an attention-grabbing user interface with animations. The chapter also touches upon the state machine framework.

Chapter 9, Testing and Debugging, explores different debugging techniques for a Qt application. You will learn about unit testing and the Qt Test framework in this chapter. We will also discuss how to use the Google Test framework with Qt Test, as well as learn about the available Qt tooling and GUI specific testing techniques.

Chapter 10, Deploying Qt Applications, discusses the importance of software deployment. You will learn to deploy a Qt application on various platforms, including desktop and mobile platforms. You will learn about the available deployment tools and steps to create an installer package.

Chapter 11, Internationalization, introduces you to internationalization. Qt provides excellent support for translating Qt Widgets and Qt Quick applications into local languages. In this chapter, you will learn how to make an application with multi-lingual support. You will also learn about inbuilt tools and various considerations for making a translation-aware application.

Chapter 12, Performance Considerations, introduces you to performance optimization techniques and how to apply them in the context of Qt programming. Here, we will discuss different profiling tools to diagnose performance problems, concentrating specifically on the tools available on Windows. In this chapter, you will learn how to profile performance with QML Profiler and benchmark your code. The chapter will also help you write high-performance optimized QML code.

To get the most out of this book

We will only use open source software, so you will not need to purchase any license. We will go over the installation procedures and detail as we progress through each chapter. To install the required software, you will require a functional internet connection and a desktop PC or laptop. Apart from that, there is no particular software requirement before you begin this book.

Main requirements for the book	OS Requirements
Qt 6.0.0 or higher. Preferably Qt 6.1.0	Windows 10 or Ubuntu 20.04 or macOS 10.14, Android 9 or above
Qt Creator 4.14.0 or higher	
Compilers such as MinGW or MSVC or GCC or LLVM	

Important notes

For Android setup, you will need the following:

OpenJDK 8 (JDK-8.0.275.1)

Android SDK 4.0

NDK r21 (21.3.6528147)

Clang toolchain

Android OpenSSL

If you are using the digital version of this book, we advise you to type the code yourself or access the code via the GitHub repository (link available in the next section). Doing so will help you avoid any potential errors related to the copying and pasting of code.

All code examples have been tested using Qt 6 on the Windows platform. You may see failures if you use Qt 5. However, they should work with future version releases too. Please make sure that the version you're installing to your computer is at least Qt 6.0.0 or later so that the code is compatible with the book.

Download the example code files

You can download the example code files for this book from GitHub at https://
github.com/PacktPublishing/Cross-Platform-Development-with-Qt-
6-and-Modern-Cpp. Additionally, you can find some bonus examples with C++17
features in the aforementioned mentioned GitHub link. In case there's an update to the
code, it will be updated on the existing GitHub repository.

We also have other code bundles from our rich catalog of books and videos available at
https://github.com/PacktPublishing/. Check them out!

Download the color images

We also provide a PDF file that has color images of the screenshots/diagrams used
in this book. You can download it here: https://static.packt-cdn.com/
downloads/9781800204584_ColorImages.pdf.

Conventions used

There are a number of text conventions used throughout this book.

Code in text: Indicates code words in text, database table names, folder names,
filenames, file extensions, pathnames, dummy URLs, user input, and Twitter handles.
Here is an example: "Typically, the exec () method is used to show a dialog."

A block of code is set as follows:

```
QMessageBox messageBox;
messageBox.setText("This is a simple QMessageBox.");
messageBox.exec();
```

When we wish to draw your attention to a particular part of a code block, the relevant
lines or items are set in bold:

```
QMessageBox messageBox;
messageBox.setText("This is a simple QMessageBox.");
messageBox.exec();
```

Any command-line input or output is written as follows:

```
> lrelease *.ts
```

Bold: Indicates a new term, an important word, or words that you see onscreen. For example, words in menus or dialog boxes appear in the text like this. Here is an example: "The last step is to build and run the application. Hit the **Run** button in Qt Creator."

> **Tips or important notes**
> Appear like this.

Get in touch

Feedback from our readers is always welcome.

General feedback: If you have questions about any aspect of this book, mention the book title in the subject of your message and email us at customercare@packtpub.com.

Errata: Although we have taken every care to ensure the accuracy of our content, mistakes do happen. If you have found a mistake in this book, we would be grateful if you would report this to us. Please visit www.packtpub.com/support/errata, selecting your book, clicking on the Errata Submission Form link, and entering the details.

Piracy: If you come across any illegal copies of our works in any form on the Internet, we would be grateful if you would provide us with the location address or website name. Please contact us at copyright@packt.com with a link to the material.

If you are interested in becoming an author: If there is a topic that you have expertise in and you are interested in either writing or contributing to a book, please visit authors.packtpub.com.

Reviews

Please leave a review. Once you have read and used this book, why not leave a review on the site that you purchased it from? Potential readers can then see and use your unbiased opinion to make purchase decisions, we at Packt can understand what you think about our products, and our authors can see your feedback on their book. Thank you!

For more information about Packt, please visit packt.com.

Section 1: The Basics

In this section, you will learn the basics and evolution of the framework and how to install Qt on different platforms. Throughout this section, you will learn more about the evolution of Qt. Then, we'll proceed to build our first example program using the latest version of Qt, which is Qt 6. You will be learning about the usage of the Qt Creator IDE. This section will introduce you to Qt Widgets, Qt Designer, and creating custom controls. You will learn about style sheets, QSS files, and theming. This section will also introduce you to Qt Quick and QML.

This section includes the following chapters:

- *Chapter 1, Introduction to Qt 6*
- *Chapter 2, Introduction to Qt Creator*
- *Chapter 3, GUI Design Using Qt Widgets*
- *Chapter 4, Qt Quick and QML*

1
Introduction to Qt 6

Qt (pronounced *cute*, not *que-tee*) is a cross-platform application development framework designed to create great software applications with uniform **user interfaces** (**UIs**) for desktop, embedded, and mobile platforms. It provides developers with a great set of tools to design and build great applications without worrying about platform dependency. In this chapter, you will learn the basics about the framework, its history, and how to install Qt on different platforms. You will learn what Qt is and why it is beneficial to use it. By the end of the chapter, you will be able to install Qt and get started on your platform of choice.

In this chapter, we're going to cover the following main topics:

- Introducing Qt
- Reasons for using Qt
- Downloading and installing Qt
- Building Qt 6 from source

Technical requirements

To get started, you should have a desktop or laptop running on Windows, Linux, or macOS. Please use the updated Windows 10 or Ubuntu 20.04 **long-term support** (**LTS**). Alternatively, use the latest version of macOS (newer than macOS 10.14), such as macOS Catalina.

For your **integrated development environment (IDE)** to run smoothly, your system should have at least an Intel Core i5 processor along with a minimum of 4 **gigabytes (GB)** of **random-access memory (RAM)**.

You will need an active internet connection to download and install Qt. As a prerequisite, you should also be familiar with C++ as Qt requires C++ programming knowledge.

Introducing Qt

Qt is a cross-platform software development framework for desktop, embedded, and mobile platforms. It follows the philosophy of *code less, create more, and deploy everywhere*. It supports platforms such as Windows, Linux, macOS, VxWorks, QNX, Android, iOS, and so on. The software also supports several **microcontroller units (MCUs)** from NXP, Renesas, and STMicroelectronics running on bare metal or FreeRTOS.

Qt was born as an attempt to provide a uniform **graphical user interface (GUI)** with the same look, feel, and functionality across different platforms. Qt accomplishes that by providing a framework to write code once and ensure that it runs on other platforms with minimal or no modifications. It is not a programming language, but rather a framework written in C++. The Qt framework and tools are dual-licensed under open source and commercial licenses.

Qt uses a modular approach to group related functionalities together. Qt Essentials are the foundation of Qt on all platforms. These modules are general and useful for most Qt-based applications. Essential modules are available for open source usage. Examples of Qt Essentials modules are Qt Core, Qt GUI, Qt QML, Qt Widgets, and so on. There are also special-purpose add-on modules that provide specific functionalities and come with certain license obligations. Examples of add-on modules are Qt 3D, Qt Bluetooth, Qt Charts, Qt Data Visualization, and more. As well as this, there are value-added modules such as Qt Automotive Suite, Qt for Device Creation, and Qt for MCUs, among others available under the commercial license.

To find out more about different Qt modules, visit `https://doc.qt.io/qt-6/qtmodules.html`.

Qt was released for public use in 1995. Since then, there have been many improvements and major changes. Qt 6 is the new major version of Qt. Its main goals are to remain prepared for the requirements coming in 2020 and beyond, remove obsolete modules, and be more maintainable. With this focus, there are architectural changes in Qt 6 that may break some level of backward compatibility with earlier versions.

Some essential modifications in Qt 6 are outlined here:

- Introduction of strong typing
- JavaScript as an optional feature of **Qt Modeling Language** (**QML**)
- Removal of QML versioning
- Removal of the duplicate data structures between QObject and QML
- Avoidance of the creation of runtime data structures
- Compilation of QML into efficient C++ and native code
- Support for hiding implementation details
- Better integration of tools

Now that we've covered the basics, let's look at the main reasons for using Qt...

Reasons for using Qt

Qt is a modular, cross-platform application development framework. The biggest misunderstanding about Qt is that many people consider it as a GUI framework. However, Qt is much more than a GUI framework. It not only comprises a GUI module, but also a set of modules to make application development faster and easier to scale on various platforms. The biggest benefit of using Qt is its ability to provide portability to various platforms. Here are some advantages of using Qt for developers:

- You can create incredible user experiences for your customers and boost your company brand using Qt.
- Cross-platform development saves both time and money. You can target multiple platforms with the same code base.
- Qt is known for making C++ easy and accessible. With Qt, developers can easily create high-performance, scalable applications with a fluid UI.
- Due to the open source model, the framework is future-proof, along with a great ecosystem.
- It further supports different programming languages and is a very flexible and reliable framework. Consequently, there are great companies such as Adobe, Microsoft, Samsung, AMD, HP, Philips, and MathWorks using Qt for their applications. Many open source projects such as VLC (previously known as VideoLAN Client), **Open Broadcaster Software** (**OBS**), and WPS Office (where **WPS** stands for **Writer, Presentation, and Spreadsheets**) are also built on Qt.

The core values of Qt are outlined as follows:

- Cross-platform nature

- Highly scalable

- Very easy to use

- Built-in with world-class **application programming interfaces (APIs)**, tools, and documentation

- Maintainable, stable, and compatible

- A large community of users

Whether you are a hobbyist, a student, or working for a company, Qt provides great flexibility to use its modules as per your requirement. Many universities are using Qt as one of their coursework subjects. So, Qt is a great choice for programmers to start building new applications with ready-made features. Let's start by downloading and installing Qt 6 on your machine.

Downloading and installing Qt

There are different ways to install the Qt framework and tools on your system. You can download an online or offline installer from the Qt website, or you can build the source packages yourself. Qt recommends using the online installer for first-time installations and the **Qt Maintenance Tool** for modifying the installation later.

The installers allow you to download and install the following components:

- Qt libraries

- Qt Creator IDE

- Documentation and examples

- Qt source code

- Add-On modules

The online installer allows you to select open source or commercial versions of Qt, tools, and Add-On modules to install based on the chosen license. The online installer doesn't contain the Qt components, but it is a downloader client to download all the relevant files. You can install once the download is complete. You will require a Qt account to download and install Qt. An evaluation version for the commercial Qt gives you free trial-period access, with all commercial packages and access to official Qt support. The installer requires you to sign in with your Qt account. If you don't have a Qt account, you can sign up during the installation process. The installer fetches the license attached to the account from the Qt server and lists down modules according to your license. If you are new to Qt, then we recommend that you start with the open source version.

The offline installer is a platform-specific package that includes all Qt modules and add-ons relevant for the platform. Due to the official policy changes, open source offline installers are not available any more since Qt 5.15. If you have a commercial license, then you can provide the credentials during the installation process. You can locate your license key in your **Qt account** web portal.

You can download them from the following links:

- **Open source**: `https://www.qt.io/download-open-source`
- **Commercial**: `https://www.qt.io/download`
- **Offline**: `https://www.qt.io/offline-installers`

> Important note
>
> The Qt Company provides users with a dual-licensing option. As a beginner, you can get started with an open source license to explore Qt. If you are working for a company, then discuss with your manager or **Information Technology** (**IT**) or legal team to procure a commercial license or to understand legal obligations. You can learn more about Qt licensing at `https://www.qt.io/licensing/`.

Downloading Qt

Let's start by downloading Qt onto your machine, as follows:

1. To begin, visit the `https://www.qt.io/download` download page.
2. Click on the **Download. Try. Buy.** button in the top-right corner. You will see different download options here.

3. If you want to try the commercial version, then click on **Try Qt** section. If you already have a Qt account, then you can log in into the account under the **Existing customers** section.

4. Considering that you are new to Qt, we will begin with the open source version. Click on the **Go open source** button, as shown in the following screenshot:

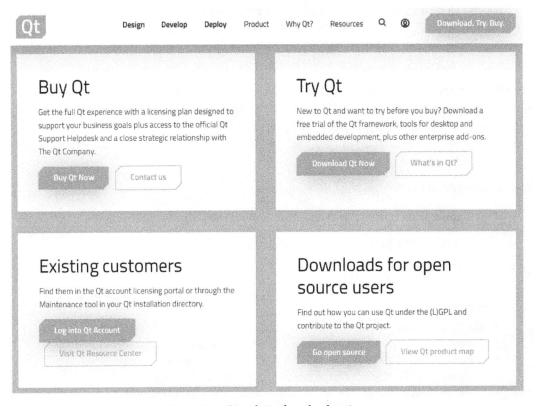

Figure 1.1 – Qt website download options

5. On the next screen, you will find **Download the Qt Online Installer** button. Click on it to proceed to the download link.

6. The web page will automatically detect the underlying platform details from the browser and will show you the **Download** button. You can also select your intended installer by choosing the other options: you can select **32-bit** or **64-bit** or download for a different platform.

    You will see a **Thank you** page after you click on the download option. At this stage, you can find the installer in your download folder.

Next, let's begin with the installation process on the Windows platform.

Installing Qt on Windows

Now, let's start the installation process on Windows! Proceed as follows:

1. You will find a file with the name `qt-unified-windows-x86-%VERSION%-online.exe` inside your download folder. Double-click on the executable, and you will see a **Welcome** screen.

2. Click the **Next** button, and a credentials screen will appear, asking you to log in with your Qt account. If you don't have one, then you can sign up on the same page, as shown in the following screenshot:

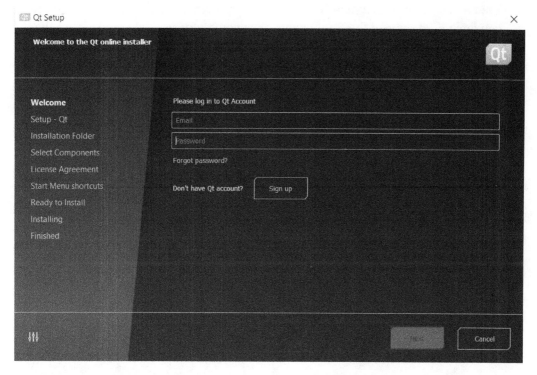

Figure 1.2 – Login screen of the installer

3. In the next screen, you will be presented with the open source usage obligations agreement. You won't get this screen if you are installing using a commercial license. Click on the first checkbox, saying **I have read and approve the obligations of using Open Source Qt**, and acknowledge that you are not using Qt for commercial purposes. Make sure you read the terms and conditions mentioned in the agreement! Then, click on the **Next** button.

4.   The next screen will provide you with options related to tracking and sharing pseudonymous data in Qt Creator. You may allow or disable these options based on your preferences. Then, click on the **Next** button to proceed to the next screen.

5.   In the next screen, you can specify the installation path. You may continue with the default path, or you can change it to any other path if you don't have enough space on the default drive. You can also choose whether you want to associate common file types with Qt Creator by selecting the checkbox option at the bottom. Click on the **Next** button.

6.   Next, you will be presented with a list where you can select the version(s) of Qt you need to install on your system. You may simply proceed with the default options. If you don't need some of the components, then you can unselect them to reduce the size of the download. You can update the Qt components using the **Maintenance Tool** anytime later. To complete the installation process, click on the **Next** button. The component selection screen can be seen here:

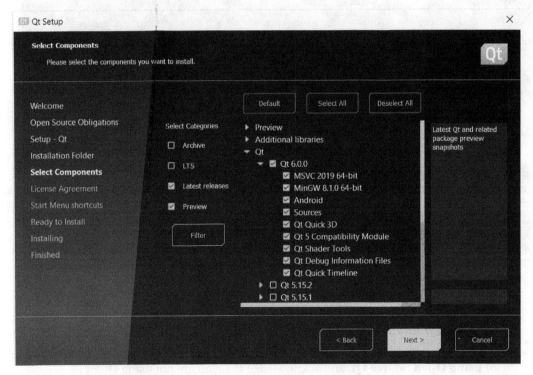

Figure 1.3 – Component selection screen of the installer

7. In the next screen, you will be presented with the license agreement. Click on the first radio button, which says **I have read and agree to the terms contained in the license agreements**. Again, make sure you read the terms and conditions mentioned in the license agreement, and then click on the **Next** button.

8. On the next screen, you can create **Start** menu shortcuts on Windows. This screen will not be available for other platforms. Once you have finished doing this, click on the **Next** button.

9. Now, Qt is ready to be installed in your system. Make sure you have a working internet connection and data balance. Click on the **Install** button to begin the installation. The download process will take time, depending on your internet speed. Once the required files are downloaded, the installer will automatically install them in the previously selected path.

10. Once the installation is finished, the installer will make an entry for the **Maintenance Tool**, which will help you later to make changes to the libraries. Click on the **Next** button to move to the last screen of the installer.

11. In order to complete the installation process, click on the **Finish** button. If you have left the **Launch Qt Creator** checkbox checked, then Qt Creator will be launched. We will discuss this in more detail in the next chapter. Now, Qt is ready to be used on your Windows machine. Click on the **Finish** button to exit the wizard.

Installing Qt on Linux

Now, let's install the Qt framework on the latest **LTS version of Linux**, such as Ubuntu 20.04, CentOS 8.1, or openSUSE 15.1. We will be focusing on the most popular Linux distribution, Ubuntu. You can follow the same steps as mentioned previously to download the online installer from the Qt website.

On Ubuntu, you will get an installer file such as `qt-unified-linux-x64-%VERSION%-online.run`, where `%VERSION%` is the latest version— for example: `qt-unified-linux-x86-4.0.1-1-online.run`.

1. You may have to give write permissions to the downloaded file before executing it. To do that, open the terminal and run the following command:

```
$ chmod +x qt-unified-linux-x64-%VERSION%-online.run
```

2. You can start the install process by double-clicking the downloaded installer file. The installation requires superuser access. You may have to add a password in the authorization dialog during the installation. You can also run the installer from the terminal, as follows:

```
$ ./qt-unified-linux-x64-%VERSION%-online.run
```

3. You will see similar screens to those shown for the Windows platform. Apart from the **operating system** (**OS**)-specific title bar changes, all the screens remain the same for installation in Ubuntu or similar Linux flavors.

At the time of writing the book, there was no Ubuntu or Debian package available for Qt 6 as the respective maintainers have stepped down. Hence, you may not get the Qt 6 package from the terminal.

Installing Qt on macOS

If you are a macOS user, then you can also install the same way as discussed for the earlier platforms. You can follow the same steps mentioned previously to download the online installer from the Qt website.

You will get an installer file such as qt-unified-mac-x64-%VERSION%-online. dmg, where %VERSION% is the latest version (such as qt-unified-mac-x64-4.0.1-1-online.dmg).

Qt has a dependency on Xcode. To install Qt on your Mac, you will need Xcode installed on your machine, otherwise, it will refuse to install. If you are an Apple developer, then your Mac may have Xcode installed. If you don't have Xcode installed on your machine, then you may proceed to install Xcode's **Command Line Tools** instead of Xcode. This will save time and storage space on your machine:

1. To begin, type the following command on the terminal:

```
$ xcode-select --install
```

2. If the terminal shows the following output, then your system is ready for the next steps:

```
xcode-select: error: command line tools are already
installed, use
"Software Update" to install updates
```

3. The next step is to install the Qt framework. Double-click on the installer file to launch the installation interface.

4. If the installer still complains that Xcode is not installed, then keep clicking **OK** until the message goes away permanently. Remember the installation path. Once the installation is finished, you are ready to use Qt on your machine.

Further instructions on Qt for macOS can be found at the following link:

```
https://doc.qt.io/qt-6/macos.html
```

Updating or removing Qt

Once Qt is installed, you can modify the components—including updating, adding, and removing components—using the **Maintenance Tool** under the installation directory. The directory structure remains the same for all desktop platforms. The installation directory contains folders and files, as shown in the following screenshot (on Windows):

Figure 1.4 – The Maintenance Tool inside the installed folder

Let's begin with the maintenance process! You can add, remove, and update modules using the **Maintenance Tool**. Proceed as follows:

1. Click on the `MaintenanceTool.exe` executable to launch the maintenance interface. Click on the **Next** button, and a credentials screen will appear, asking you to log in with your Qt account. The login details will be prefilled from your last login session. You can click **Next** to add or update components or select the **Uninstall only** checkbox to remove Qt from your system. The following screenshot shows what the credentials screen looks like:

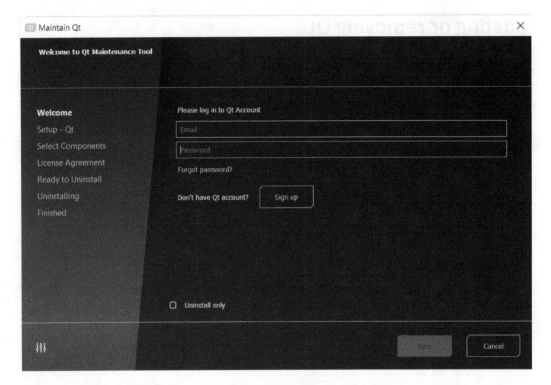

Figure 1.5 – Welcome screen of the Maintenance Tool

2. Once you are logged in, the tool will present you with options to add or remove or update the components, as shown in the following screenshot. Click on the **Next** button to proceed further:

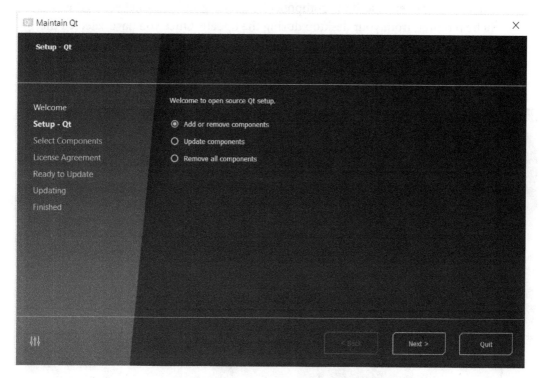

Figure 1.6 – Setup screen of the Maintenance Tool

3. On the next screen, you can select new components from the latest releases or the archived version. You can click on the **Filter** button to filter versions as per your requirement. You can also add new platform-specific components such as Android from the component list. If the component is existing and you uncheck it, then it will be removed from your desktop during the update. Once you have selected the components, click on the **Next** button. The following screenshot shows what the component selection screen looks like:

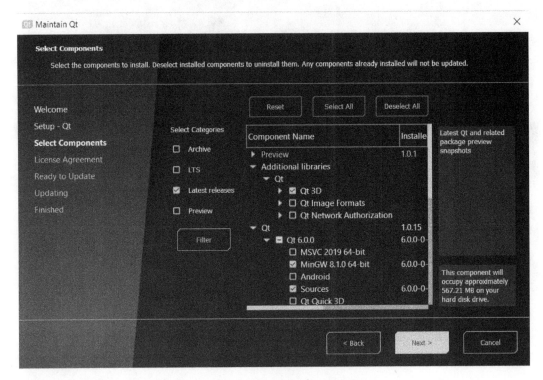

Figure 1.7 – Component selection screen

4. You will then come across the update screen. This screen will tell you how much storage space is required for the installation. If you are running out of storage space, then you may go back and remove some existing components. Click on the **Update** button to begin the process, as illustrated in the following screenshot:

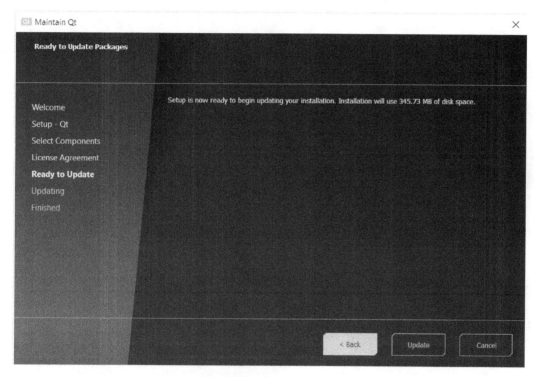

Figure 1.8 – The Ready to Update screen of the Maintenance Tool

5. You can abort the update installation process by clicking on the **Cancel** button. Qt will warn you and ask you for confirmation before aborting the installation process, as illustrated in the following screenshot. Once the process is aborted, click on the **Next** button to exit the wizard:

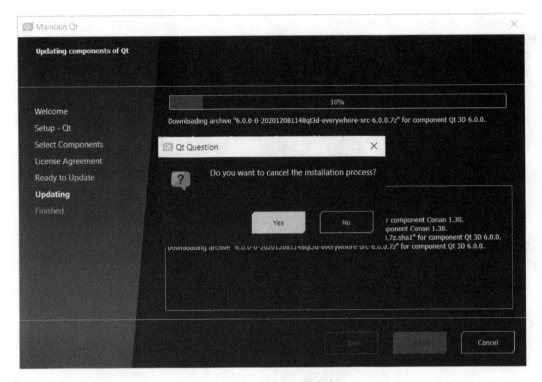

Figure 1.9 – The cancel dialog

6. Launch the **Maintenance Tool** again to update existing components from the latest releases. You can click on the **Quit** button to exit the **Maintenance Tool**. Please wait while the installer fetches the meta-information from the remote repository. Click on the **Next** button to see the available components. The update option is illustrated in the following screenshot:

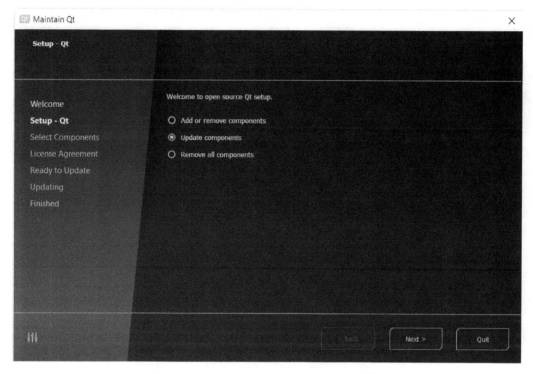

Figure 1.10 – The Update option in the Maintenance Tool

7. Next, you can select which components to update from the checkboxes. You can choose to update all or you can update selectively. The installer will show how much storage space will be required for the update, as illustrated in the following screenshot. You can click **Next** to go to the update screen and begin the update. Then, on the next screen, click on the **Update** button to download the update packages:

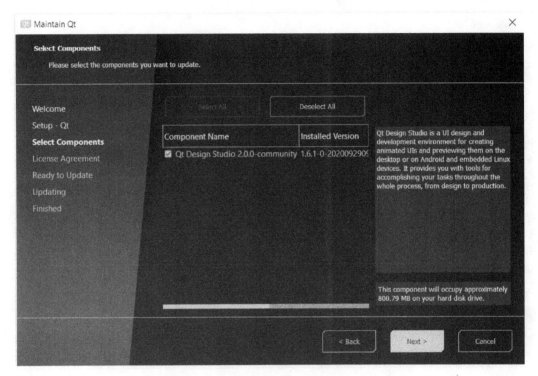

Figure 1.11 – Components available for update

8. Once the installation is finished, the installer makes entries for the Maintenance Tool, which will help you make changes to the libraries later. This is illustrated in the following screenshot. Click on the **Next** button to move to the last screen of the installer:

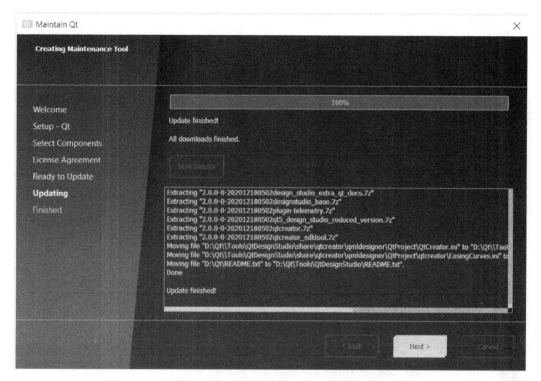

Figure 1.12 – The Update finished screen in the Maintenance Tool

9. In the last screen, you will see **Restart** and **Finish** buttons. Click on the **Finish** button to exit the Qt wizard.

10. Similarly, you can restart or launch the **Maintenance Tool** and select the **Remove all components** radio button. Click on the **Next** button to begin the uninstallation process, as illustrated in the following screenshot:

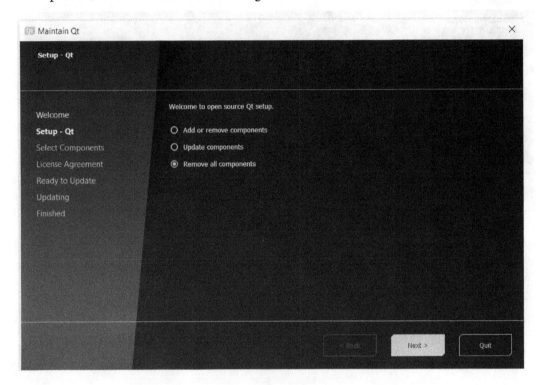

Figure 1.13 – The Remove option in the Maintenance Tool

Please note that on clicking the **Uninstall** button, all the Qt components will be removed from your system; you will have to reinstall Qt if you want to use them again. Click on **Cancel** if you don't intend to remove the Qt components from your system, as illustrated in the following screenshot. If you intend to remove the existing version and use a newer version of Qt, then select the **Add or remove components** option, as discussed earlier. This will remove older Qt modules and free up your disk space:

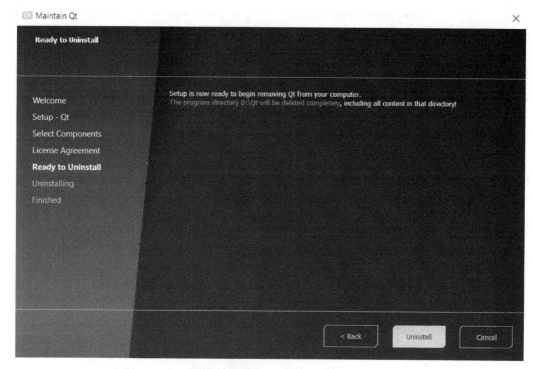

Figure 1.14 – The Uninstall screen in the Maintenance Tool

In this section, we learned about modifying an existing Qt installation through the Maintenance Tool. Now, let's learn how to build and install Qt from the source code.

Building Qt 6 from source

If you want to build the framework and tools yourself or experiment with the latest unreleased code, then you can **build Qt from the source code**. If you're going to develop a specific Qt version from the source, then you can download the Qt 6 source code from the official releases link, as shown here: `https://download.qt.io/official_releases/qt/6.0/`.

If you are a commercial customer, then you can download the **Source Packages** from your Qt account portal. Platform-specific building instructions are discussed in the upcoming subsections.

You can also clone from the GitHub repository, and check out the desired branch. At the time of authoring this book, the Qt 6 branch remained inside the Qt 5 super module. You can clone the repository from the following link: `git://code.qt.io/qt/qt5.git`.

The `qt5.git` repository may get renamed to `qt.git` in the future for maintainability. Please refer to the `QTQAINFRA-4200` Qt ticket. Detailed instructions on how to build Qt from Git can be found at the following link: `https://wiki.qt.io/Building_Qt_6_from_Git`.

Ensure that you install the latest versions of Git, Perl, and Python on your machine. Make sure there is a working C++ compiler before proceeding to the platform-specific instructions in the next section.

Installing Qt on Windows from source

To install Qt 6 on Windows from source code, follow these next steps:

1. First of all, download the source code from Git or from the open source download link mentioned earlier. You will get a compressed file as `qt-everywhere-src--%VERSION%.zip`, where `%VERSION%` is the latest version (such as `qt-everywhere-src-6.0.3.zip`). Please note that suffixes such as `-everywhere-src-` may get removed in the future.

2. Once you have downloaded the source archive, extract it to a desired directory—for example, `C:\Qt6\src`.

3. In the next step, configure the build environment with a supported compiler and the required build tools.

4. Then, add the respective installation directories of `CMake`, `ninja`, `Perl`, and `Python` to your `PATH` environment variable.

5. The next step is to build the Qt library. To configure the Qt library for your machine type, run the `configure.bat` script in the source directory.

6. In this step, build Qt by typing the following command in Command Prompt:

   ```
   >cmake --build . -parallel
   ```

7. Next, enter the following command in Command Prompt to install Qt on your machine:

   ```
   >cmake --install .
   ```

Your Windows machine is now ready to use Qt.

To understand more about the configure options, visit the following link:

`https://doc.qt.io/qt-6/configure-options.html`

Detailed build instructions can be found at the following link:

`https://doc.qt.io/qt-6/windows-building.html`

Installing Qt on Linux from source

To build the source package on Linux distributions, run the following set of instructions on your terminal:

1. First of all, download the source code from Git or from the open source download link mentioned earlier. You will get a compressed file as `qt-everywhere-src--%VERSION%.tar.xz`, where `%VERSION%` is the latest version (such as `qt-everywhere-src-6.0.3.tar.xz`). Please note that suffixes such as `-everywhere-src-` may get removed in the future.

2. Once you have downloaded the source archive, uncompress the archive and unpack it to a desired directory—for example, `/qt6`, as illustrated in the following code snippet:

   ```
   $ cd /qt6
   $ tar xvf qt-everywhere-opensource-src-%VERSION%.tar.xz
   $ cd /qt6/qt-everywhere-opensource-src-%VERSION%
   ```

3. To configure the Qt library for your machine, run the `./configure` script in the source directory, as illustrated in the following code snippet:

   ```
   $ ./configure
   ```

4. To create the library and compile all the examples, tools, and tutorials, type the following commands:

   ```
   $ cmake --build . --parallel
   $ cmake --install .
   ```

5. The next step is to set the environment variables. In `.profile` (if your shell is bash, ksh, zsh, or sh), add the following lines of code:

   ```
   PATH=/usr/local/Qt-%VERSION%/bin:$PATH
   export PATH
   ```

   In `.login` (if your shell is csh or tcsh), add the following line of code:

   ```
   setenv PATH /usr/local/Qt-%VERSION%/bin:$PATH
   ```

If you use a different shell, modify your environment variables accordingly. Qt is now ready to be used on your Linux machine.

Detailed building instructions for Linux/X11 can be found at the following link:

```
https://doc.qt.io/qt-6/linux-building.html
```

Installing Qt on macOS from source

Qt has a dependency on **Xcode**. To install Qt on your Mac, you will need Xcode installed on your machine. If you don't have Xcode installed on your machine, then you may proceed to install Xcode's **Command Line Tools**:

1. To begin, type the following command on the terminal:

    ```
    $ xcode-select --install
    ```

2. If the terminal shows the following output, then your system is ready for the next steps:

    ```
    xcode-select: error: command line tools are already
    installed, use
    "Software Update" to install updates
    ```

3. To build the source package, run the following set of instructions on your terminal:

    ```
    $ cd /qt6
    $ tar xvf qt-everywhere-opensource-src-%VERSION%.tar
    $ cd /qt6/qt-everywhere-opensource-src-%VERSION%
    ```

4. To configure the Qt library for your Mac, run the ./configure script in the source directory, as illustrated in the following code snippet:

    ```
    $ ./configure
    ```

5. To create a library, run the make command, as follows:

    ```
    $ make
    ```

6. If -prefix is outside the build directory, then type the following lines to install the library:

    ```
    $ sudo make -j1 install
    ```

7. The next step is to set the environment variables. In `.profile` (if your shell is bash), add the following lines of code:

```
PATH=/usr/local/Qt-%VERSION%/bin:$PATH
export PATH
```

In `.login` (if your shell is csh or tcsh), add the following line of code:

```
setenv PATH /usr/local/Qt-%VERSION%/bin:$PATH
```

Your machine is now ready for Qt programming.

Detailed building instructions for macOS can be found here:

`https://doc.qt.io/qt-6/macos-building.html`

In this section, we learned how to install Qt from source on your favorite platform. Now, let's summarize our learning.

Summary

This chapter explained the basics of the Qt framework and what it can be used for. Here, we discussed the history, different modules, and advantages of using Qt. We also learned about different methods of installation with license obligations, giving step-by-step installation procedures for Qt on different desktop platforms. Now, your machine is ready to explore Qt.

In the next chapter, we will discuss the Qt Creator IDE. You will learn about the UI of the IDE, different configurations, and how to use it for your Qt project.

2
Introduction to Qt Creator

Qt Creator is Qt's own **Integrated Development Environment** (**IDE**) for cross-platform application development. In this chapter, you will learn the basics of the Qt Creator IDE as well as covering the **user interface** (**UI**) of the IDE. We will also look at how to create and manage projects in Qt Creator. This module of Qt covers developing a simple Qt application using Qt Creator, shortcuts, and practical tips for developers.

More specifically, we're going to cover the following main topics:

- Basics of Qt Creator
- Configuring the IDE and managing projects
- User interfaces
- Writing a sample application
- Advanced options

Qt Creator can make your Qt learning easier with many useful tools and examples. You will need minimal IDE knowledge to get started. By the end of this chapter, you will be familiar with the use of Qt Creator. You will also be able to build and run your first Qt application on your favorite desktop platform, as well as being aware of the advanced options available in the IDE, which you will be able to customize in line with your preferences.

Technical requirements

The technical requirements for this chapter are the same as *Chapter 1, Introduction to Qt 6*. You will need the latest Qt version, namely Qt 6.0.0 MinGW 64-bit, Qt Creator 4.13.0 or higher, and Windows 10, Ubuntu 20.04 LTS, or the latest version of macOS (higher than macOS 10.13 at a minimum) such as macOS Catalina. Qt supports earlier versions of operating systems such as Windows 8.1 or Ubuntu 18.04. However, we recommend you upgrade to the latest version of your preferred operating system to ensure smooth functioning. In this chapter, we have used screenshots from the Windows 10 platform.

Exploring the Qt Creator UI

Qt Creator is an IDE produced by the Qt Company. It integrates multiple tools including a code editor, a **Graphical UI (GUI)** designer, a compiler, a debugger, Qt Designer, Qt Quick Designer, and Qt Assistant, among others.

Qt Designer helps in designing widget-based GUIs whereas Qt Quick Designer provides a UI to create and edit QML-based GUIs in Design Mode. Qt Assistant is an integrated documentation viewer that opens contents related to a given Qt class or function with the press of the *F1* key.

Let's begin by launching Qt Creator. The binary can be found inside `Qt\Tools\ QtCreator\bin`. You will see a screen like that shown in *Figure 2.1*:

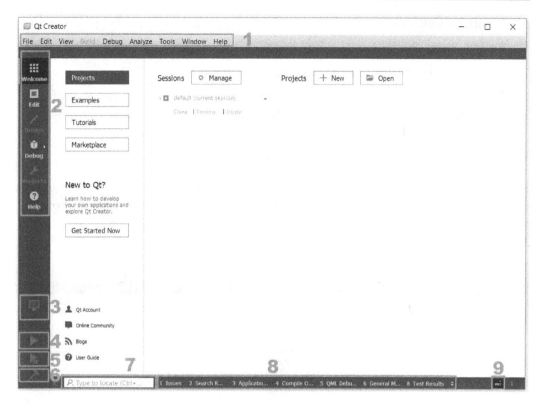

Figure 2.1 – Qt Creator interface

You can see the following GUI sections in the UI:

1. **IDE menu bar**: This provides the user with a standard place in the window to find the majority of application-specific functions. These functions include creating a project, opening and closing files, development tools, analysis options, help contents, and a way to exit the program.

2. **Mode selector**: This section provides different modes depending on the active task. The **Welcome** button gives options to open examples, tutorials, recent sessions, and projects. The **Edit** button opens the code window and helps in navigating the project. The **Design** button opens Qt Designer or Qt Quick Designer based on the type of UI file. **Debug** provides options to analyze your application. The **Projects** button helps in managing project settings, and the **Help** button is for browsing help contents.

3. **Kit selector**: This helps in selecting the active project configuration and changing the kit settings.

4. **Run button**: This button runs the active project after building it.

5. **Debug button**: This helps in debugging the active project using a debugger.

6. **Build button**: This button is for building the active project.

7. **Locator**: This is used to open a file from any open project.

8. **Output pane**: This includes several windows to display project information such as compilation and application output. It also shows build issues, console messages, and test and search results.

9. **Progress indicator**: This control shows the progress related to running tasks.

You can also benefit from an interactive UI tour when you launch Qt Creator for the first time. You can also launch it from the **Help | UI Tour** option from the menu bar as shown in *Figure 2.2*:

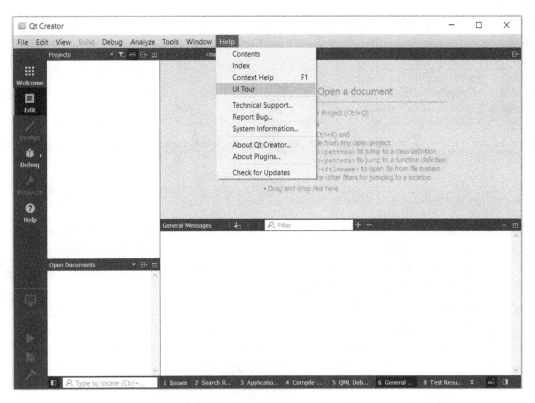

Figure 2.2 – Qt Creator UI Tour menu choice

> **Note**
> If you press the *Alt* key, then you will see the underlined mnemonic letter in the menu title. Press the corresponding key to open the respective context menu.

In this section, we learned about various sections in the IDE. In the next section, we will build a simple Qt application using the Qt Creator IDE.

Building a simple Qt application

Let's start with a simple *Hello World* project. A *Hello World* program is a very simple program that displays **Hello World!** and checks that the SDK configuration is free from errors. These projects use the most basic, very minimal code. For this project, we will use a project skeleton created by Qt Creator.

Follow these steps to build your first Qt application:

1. To create a new project in Qt, click on the **File** menu option on the menu bar or hit *Ctrl + N*. Alternatively, you can also click on the **+ New** button located at the welcome screen to create a new project, as shown in *Figure 2.3*:

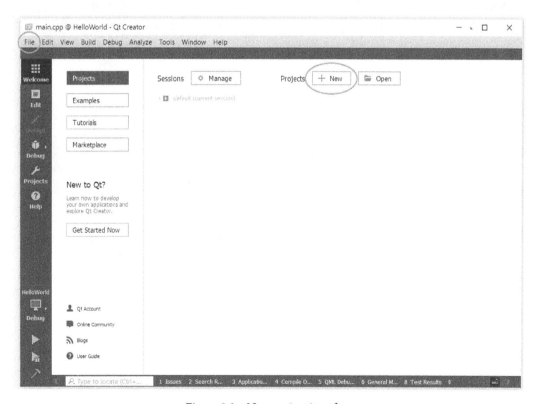

Figure 2.3 – New project interface

2. Next, you can select a template for the project. You can create different types of applications, including a console application or GUI application. You can also create non-Qt projects as well as library projects. In the upper-right section, you will see a dropdown to filter templates specific to the desired target platform. Select the **Qt Widgets Application** template and then click on the **Choose...** button:

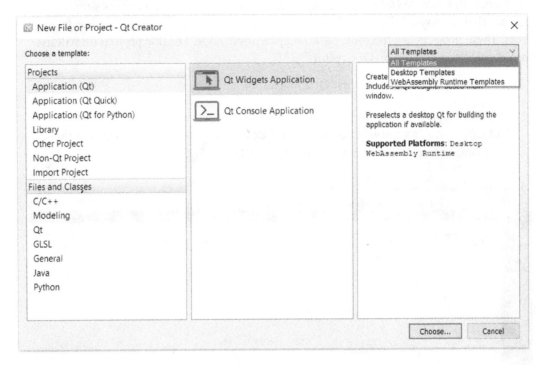

Figure 2.4 – The project template interface

3. In the next step, you will be asked to choose the project name and project location. You can navigate to the desired project location by clicking the **Browse...** button. Then click on the **Next** button to proceed to the next screen:

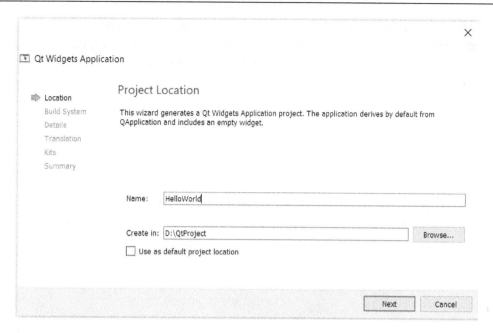

Figure 2.5 – New project location screen

4. You can now select the build system. By default, Qt's own build system **qmake** will be selected. We will discuss qmake more later in *Chapter 6, Signals and Slots*. Click on the **Next** button to proceed to the next screen:

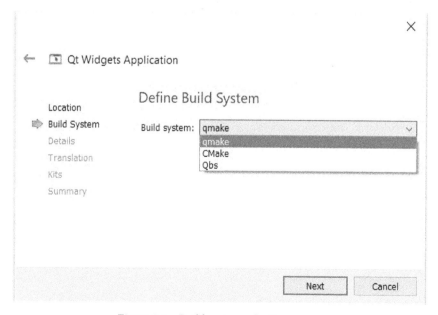

Figure 2.6 – Build system selection screen

5. Next, you can specify the class information and the base class to be used to automatically generate the project skeleton. If you need a desktop application with features of MainWindow such as menubar, toolbar, and statusbar, then select **QMainWindow** as the base class. We will discuss more on QMainWindow in *Chapter 3, GUI Design Using Qt Widgets.* Click on the **Next** button to proceed to the next screen:

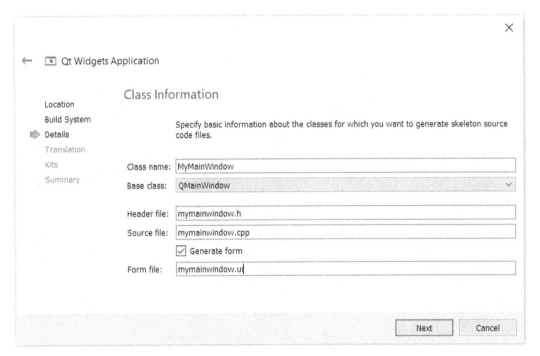

Figure 2.7 – Source code skeleton generation screen

6. In this next step, you can specify the language for translation. Qt Creator comes with the *Qt Linguist* tool, which allows you to translate your application into different languages. You may skip this step for now. We will discuss **Internationalization (i18n)** in *Chapter 11, Internationalization*. Click on the **Next** button to proceed to the next screen:

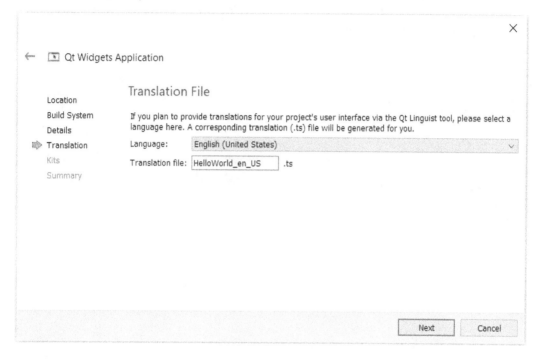

Figure 2.8 – Translation File creation screen

7. In the next step, you can select a kit to build and run your project. To build and run the project, at least one kit must be active and selectable. If your desired kit is shown as grayed out, then you may have some kit configuration issues. When you install Qt for a target platform, the build and run settings for the development targets usually get configured automatically. Click on the checkbox to select one of the desktop kits such as **Desktop Qt 6.0.0 MinGW 64-bit**. Click on the **Next** button to proceed to the next screen:

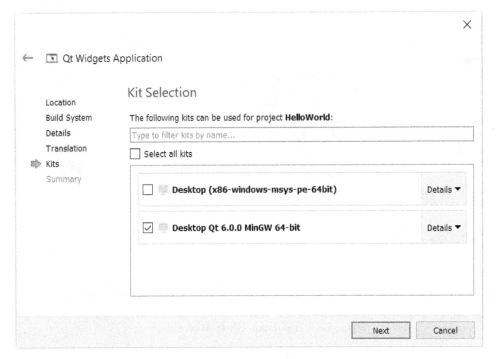

Figure 2.9 – Kit Selection screen

8. Version control allows you or your team to submit code changes to a centralized system so that each and every team member can obtain the same code without passing files around manually. You can add your project into the installed version control system on your machine. Qt has support for several version control systems within the Qt Creator IDE. You may skip version control for this project by selecting **<None>**. Click on the **Finish** button to complete the project creation:

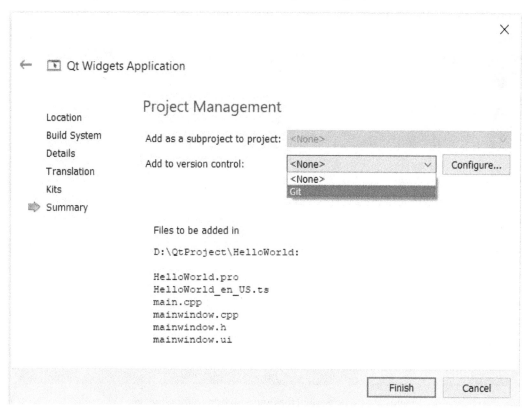

Figure 2.10 – Project management screen

9. Now you will see the generated files on the left side of the editor window. Click on any file to open it in the coding window, the most used component of the Qt Creator. The code editor is used in **Edit** mode. You can write, edit, refactor, and beautify your code in this window. You can also modify the fonts, font size, colors, and indentation. We will learn more about these in the *Understanding advanced options* section later in this chapter:

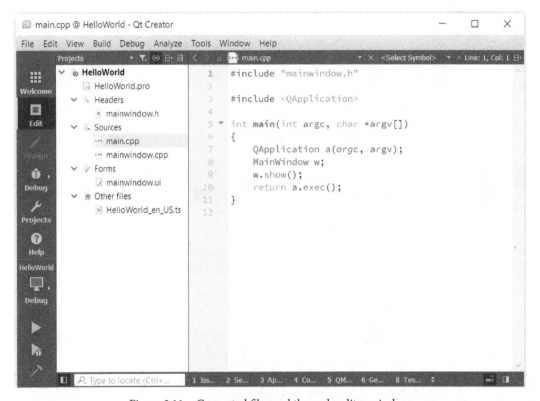

Figure 2.11 – Generated files and the code editor window

10. You can now see a `.pro` file inside your project folder. In the current project, the `HelloWorld.pro` file is the project file. This contains all the information required by qmake to build the application. This file is autogenerated during the project creation and contains the relevant details in a structured fashion. You can specify files, resources, and target platforms in this file. You need to run qmake again if you make any modifications to the `.pro` file contents, as shown in *Figure 2.12*. Let's skip modifying the contents for this project:

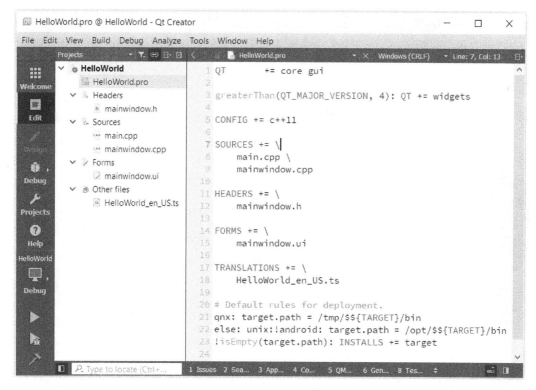

Figure 2.12 – The contents of the project file

11. You can find a form file with the `.ui` extension in the left side of the editor window. Let's open the `mainwindow.ui` file by double-clicking it. Here, you can see the file opens under a different interface: Qt Designer. You can see that the mode selection panel has switched to **Design** mode. We will discuss Qt Designer more in the next chapter.

12. Now drag the **Label** control listed under the **Display Widgets** category to the center of the form on the right side, as shown in *Figure 2.13*.

13. Next, double-click on the item you dragged in, and type `Hello World!`. Hit the *Enter* key on your keyboard or click with the mouse anywhere outside the control to save the text:

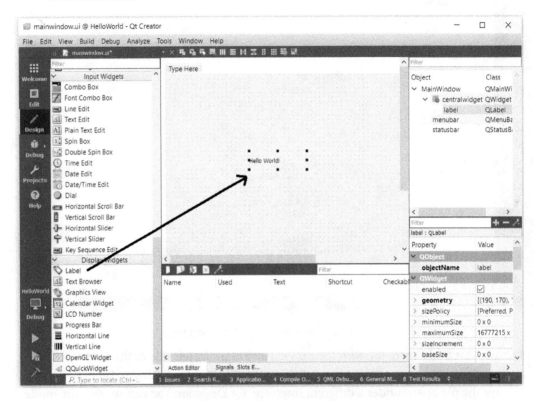

Figure 2.13 – The designer screen

14. The final step is to press the **Run** button present below the kit selector button. The project will build automatically once the reader hits the **Run** button. Qt Creator is intelligent enough to figure out that the project needs to be built first. You can build and run the application separately. After a few seconds of compiling, you will see a window with text reading **Hello World!**:

Figure 2.14 – The display output of the sample GUI application

Congratulations, you have created your first Qt-based GUI application! Now let's explore the different advanced options available in Qt Creator.

Understanding advanced options

When you install Qt Creator, it gets installed with the default configuration. You can customize the IDE and configure its look and feel or set your favorite coding style.

Go to the top menu bar and click on the **Tools** option, then select **Options…**. You will see a list of categories available on the left sidebar. Each category provides a set of options to customize Qt Creator. As a beginner, you may not need to change the settings at all, but let's get familiar with the different options available. We will start by looking at managing kits.

Managing kits

Qt Creator can automatically detect the installed Qt versions and available compilers. It groups the configurations used for building and running projects to make them cross-platform compatible. This set of configurations are stored as a kit. Each kit contains a set of parameters that describe the environment, such as the target platform, compiler, and Qt version.

Start by clicking on the **Kits** option in the left sidebar. This will autodetect and list the available kits as shown in *Figure 2.15*. If any kit is shown with a yellow or red warning mark, then it signifies a fault in the configuration. In that case, you may have to select the right compiler and Qt version. You can also create a customized kit by clicking on the **Add** button. If you want to use a new kit, then don't forget to click on the **Apply** button. We will proceed with the default desktop configuration shown as follows:

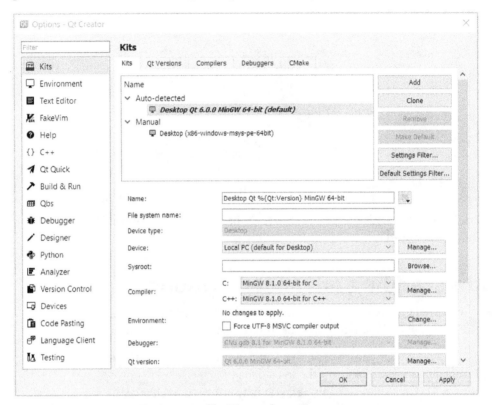

Figure 2.15 – The Kits configuration screen

Now let's proceed to the **Qt Versions** tab under the **Kits** section.

Qt Versions

In this tab, you can see the Qt versions available on your system. Ideally, the version gets detected automatically. If it is not detected, then click on the **Add...** button and browse to the path of qmake to add the desired Qt version. Qt uses a defined numbering scheme for its releases. For example, Qt 6.0.0 signifies the first patch release of Qt 6.0 and 6 as the major Qt version. Each release has limitations on the acceptable amount of changes to ensure a stable API. Qt tries to maintain compatibility between versions. However, this is not always possible due to code clean-ups and architectural changes in major versions:

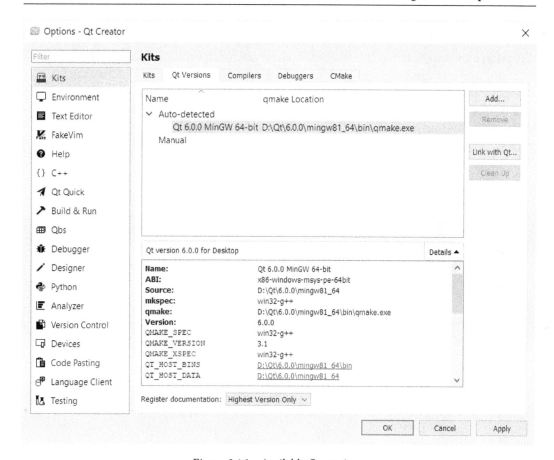

Figure 2.16 – Available Qt versions

> **Important note**
>
> Qt software versions use the versioning format of `Major.Minor.Patch`. Major releases may break backward compatibility for both the binary and source, although source compatibility may be maintained. Minor releases have binary and source backward compatibility. Patch releases have both backward and forward compatibility for the binary and the source.

We won't be discussing all of the tabs under the **Kits** section as the other tabs require knowledge about compilers, debuggers, and build systems. If you are an experienced developer, you may explore the tabs and make changes as required. Let's proceed to the **Environment** category in the left sidebar.

Environment

This option allows the user to choose their preferred language and theme. By default, Qt Creator uses the system language. It doesn't support many languages, but most of the popular languages are available. If you change to a different language, then click on the **Apply** button and restart Qt Creator to see the changes. Please note that these **Environment** options are different from the build environment. You will see an interface similar to *Figure 2.17* shown as follows:

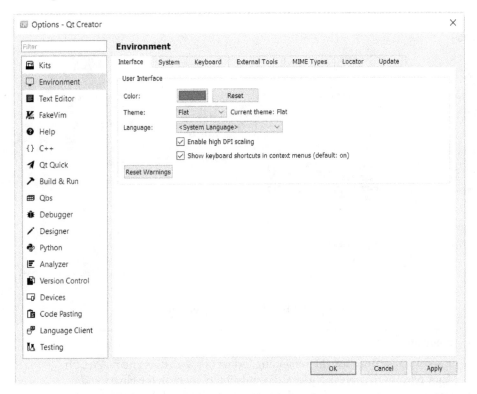

Figure 2.17 – Options for the Environment settings

You will also see a checkbox saying **Enable high DPI scaling**. Qt Creator handles high **Dots-Per-Inch** (**DPI**) scaling differently on different operating systems, as follows:

- On Windows, Qt Creator detects the default scaling factor and uses it accordingly.

- On Linux, Qt Creator leaves the decision of whether or not to enable high DPI scaling up to the user. This is done because there are many Linux flavors and windowing systems.

- On macOS, Qt Creator forces Qt to use the system scaling factor for the Qt Creator scaling factor.

To override the default approach, you may toggle the checkbox option and click the **Apply** button. The changes will be reflected after you restart the IDE. Now let's have a look at the **Keyboard** tab.

Keyboard shortcuts

The **Keyboard** section allows users to explore existing keyboard shortcuts and create new ones. Qt Creator has many built-in keyboard shortcuts, which are very useful for developers. You can also create your own shortcuts if your favorite shortcut is missing. You can additionally specify your own keyboard shortcuts for the functions that do not appear in the list, such as, for example, selecting words or lines in a text editor.

Some of the commonly used shortcuts for everyday development are listed as follows:

New File or Project	CTRL + N
Open File or Project	CTRL + O
Save Current File	CTRL + S
Save All	CTRL + SHIFT + S
Close Current File	CTRL + W
Close All	CTRL + SHIFT + W
Quit Qt Creator	CTRL + Q
Return to Edit mode	ESCAPE
Switch Current File	CTRL + TAB
Force Code Completion	CTRL + SPACE
Start Debugging	F5
Stop Debugging	SHIFT + F5
Step Over	F10
Step Into	F11
Step Out	SHIFT + F11
Toggle Breakpoint	F9
Build Current Project	CTRL + B
Run	CTRL + R
Build All	CTRL + SHIFT + B
Locate	CTRL + K

Figure 2.18 – Some of the commonly used keyboard shortcuts

The shortcuts are grouped by category. To find a keyboard shortcut in the list, enter a function name or shortcut in the **Filter** field. In *Figure 2.19*, we have searched for the available shortcuts related to new:

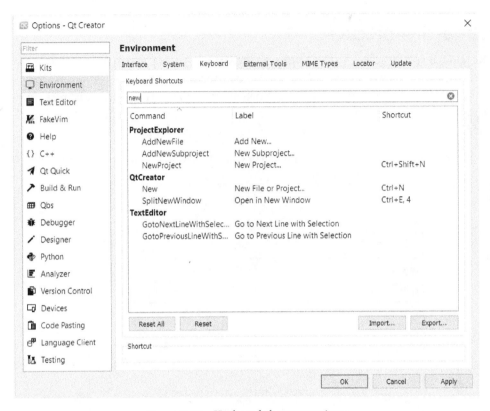

Figure 2.19 – Keyboard shortcut options

The preceding screenshot shows a list of available shortcuts for the keyword new. You can see *Ctrl + N* is used for creating a new file or project. You can also import or export keyboard mapping schemes files in .kms format.

> **Important note**
>
> There are many more in-built Qt shortcuts than we discussed here. You can read more about shortcuts in the following articles:
>
> https://doc.qt.io/qtcreator/creator-keyboard-shortcuts.html
>
> https://wiki.qt.io/Qt_Creator_Keyboard_Shortcuts
>
> https://shortcutworld.com/Qt-Creator/win/Qt-Creator_Shortcuts

There is a possibility of conflict between a Qt Creator keyboard shortcut and a window manager shortcut. In this case, the window manager shortcut will override the Qt Creator shortcut. You can also configure the keyboard shortcuts in the window manager. If this is restricted, then you can change the Qt Creator shortcuts instead. Now, let's proceed to the next sidebar category.

Text Editor

The next category in the left sidebar is **Text Editor**. Here, you can choose the color scheme, font, and font size in the first tab. The next tab lists different behavior in **Text Editor**. As you can see in *Figure 2.20*, Qt uses space indentation for the *Tab* key on the keyboard:

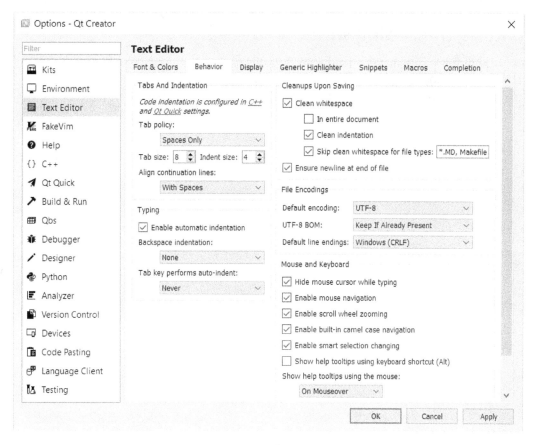

Figure 2.20 – Text Editor Behavior tab

Some developers prefer tab indentation over space indentation. You can change the indentation behavior in the **C++** and **Qt Quick** settings. Since there are dedicated settings as a different sidebar category, this section in **Text Editor** may be deprecated in future releases.

You can find the file encoding of the current file in the **File Encodings** group. To modify the file encoding, select **New Encoding** from the dropdown. To view the file with the new encoding, click on the **Apply** button.

We won't be discussing all of the sidebar categories as those are very advanced options. You can explore them later once you learn the basics. In the next section, we will discuss managing the coding window.

Splitting the coding window

You can split the coding window and view multiple files on the same screen or on an external screen. You can view multiple files simultaneously in a selection of different ways (the options are available under the **Window** option in the menu bar):

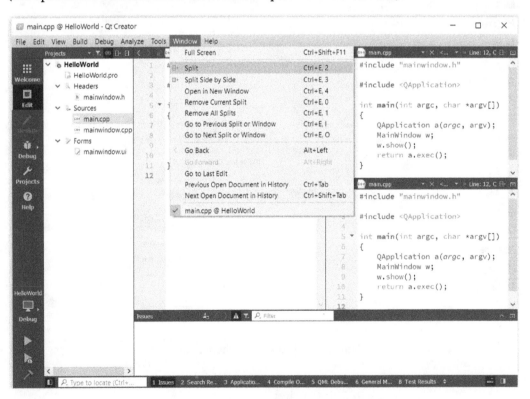

Figure 2.21– A screenshot showing the spilt screen options

Now let's discuss various ways to split a coding window and remove a split window:

- To split the coding window into top and bottom views, press *Ctrl + E* and then *2*, or select the **Window** option in the menu bar and then click on the **Split** option. This will create an additional coding window below the currently active window.

- To split the coding window into adjacent views, select **Split Side by Side** or press *Ctrl + E* and then *3*. A side-by-side split creates views to the right of the currently active coding window.

- To open the coding window in a detached window, press *Ctrl + E*, and *4*, or select **Open in New Window**. You can drag the window to an external monitor for convenience.

- To move between split views and detached editor windows, select **Next Split** or press *Ctrl + E*, and then *O*.

- To remove a split view, click on the window you want to remove and select **Remove Current Split**, or press *Ctrl + E*, and then *0*.

- To remove all split coding windows, select **Remove All Splits** or press *Ctrl + E*, and then *1*.

In this section, you learned about splitting the coding editor window. This helps when referring to multiple code files at once while coding. In the next section, we will discuss the **Build** menu present in the IDE's menu bar.

Build options

In the menu bar, you can see the **Build** option. If you click that, then you will see various build options as shown in *Figure 2.22*. Here, you can build, rebuild, or clean your projects. In complex projects, you may have more than one subproject. You can build subprojects individually to reduce total build time:

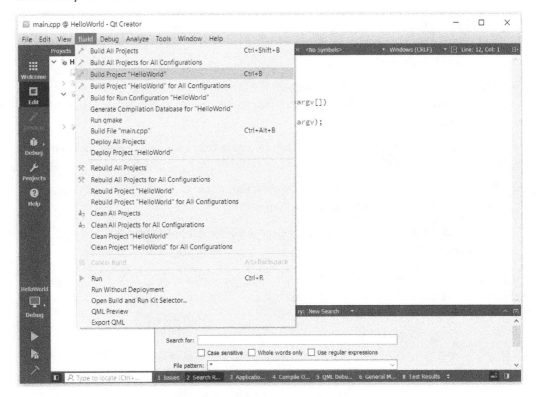

Figure 2.22 – Build menu options

Qt Creator project wizards allow you to choose the build system, including qmake, CMake, and Qbs, while creating a new project. It gives developers the freedom to use Qt Creator as a code editor and to have control of the steps or commands used in building a project. By default, qmake is installed and configured for your new project. You can learn more about using other build systems at the following link: `https://doc.qt.io/qtcreator/creator-project-other.html`.

Now let's discuss where and how to look for the framework's documentation.

Qt Assistant

Qt Creator also includes a built-in documentation viewer called Qt Assistant. This is really handy since you can look for an explanation of a certain Qt class or function by simply hovering the mouse cursor over the class name in your source code and pressing the *F1* key. Qt Assistant will then be opened and will show you the documentation related to that Qt class or function:

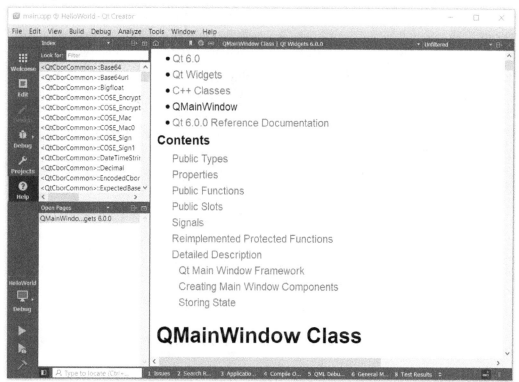

Figure 2.23 – Integrated help interface

Qt Assistant also provides support for interactive help and enables you to create help documentation for your Qt application.

> **Note**
>
> On the Windows platform, Qt Assistant is available as a menu option on the Qt Creator menu bar. On Linux distributions, you can open Terminal, type `assistant`, and press *Enter*. On macOS, it is installed in the `/Developer/Applications/Qt` directory.

In this section, we learned about Qt Assistant and the help documentation. Now, let's summarize our takeaways from this chapter.

Summary

This chapter explained the fundamentals of the Qt Creator IDE and what it can be used for. Qt Creator is an IDE with a great set of tools. It helps you to easily create great GUI applications for multiple platforms. Developers don't need to write lengthy code just to create a simple button or change a lot of code just to align a text label – Qt Designer automatically generates code for us when we design our GUI. We created a GUI application with just a few clicks, and we also learned about the various advanced options available in the IDE, including how to manage kits and shortcuts. The built-in Qt Assistant provides great help with useful examples and can help us with our own documentation.

In the next chapter, we will discuss GUI design using Qt Widgets. Here, you will learn about different widgets, how to create your own GUI element, and how to create a custom GUI application.

3

GUI Design Using Qt Widgets

Qt Widgets is a module that offers a set of **user interface** (UI) elements for building classic UIs. In this chapter, you will be introduced to the **Qt Widgets** module and will learn about basic widgets. We will look at what widgets are and the various kinds that are available for creating **graphical UIs** (**GUIs**). In addition to this, you will be introduced to layouts with **Qt Designer**, and you will also learn how to create your own custom controls. We will take a close look into what Qt can offer us when it comes to designing sleek-looking GUIs with ease. At the beginning of this chapter, you will be introduced to the types of widgets provided by Qt and their functionalities. After that, we will walk through a series of steps and design our first form application using Qt. You will then learn about Style Sheets, **Qt Style Sheets** (**QSS files**), and theming.

The following main topics will be covered in this chapter:

- Introducing Qt widgets
- Creating a UI with Qt Designer
- Managing layouts
- Creating custom widgets
- Creating Qt Style Sheets and custom themes
- Exploring custom styles
- Using widgets, windows, and dialogs

By the end of this chapter, you will understand the basics of GUI elements and their corresponding C++ classes, how to create your own UI without writing a single line of code, and how to customize the look and feel of your UI using Style Sheets.

Technical requirements

The technical requirements for this chapter include Qt 6.0.0 MinGW 64-bit, Qt Creator 4.14.0, and Windows 10/Ubuntu 20.04/macOS 10.14. All the code used in this chapter can be downloaded from the following GitHub link: `https://github.com/PacktPublishing/Cross-Platform-Development-with-Qt-6-and-Modern-Cpp/tree/master/Chapter03`.

> **Note**
> The screenshots used in this chapter are taken from a Windows environment. You will see similar screens based on the underlying platforms in your machine.

Introducing Qt widgets

A widget is the basic element of a GUI. It is also known as a **UI control**. It accepts different user events such as mouse and keyboard events (and other events) from the underlying platform. We create UIs using different widgets. There was a time when all GUI controls were written from scratch. Qt widgets reduce time by developing a desktop GUI with ready-to-use GUI controls, and Qt widely uses the concept of inheritance. All widgets inherit from QObject. QWidget is a basic widget and is the base class of all UI widgets. It contains most of the properties required to describe a widget, along with properties such as geometry, color, mouse, keyboard behavior, tooltips, and so on. Let's have a look at QWidget inheritance hierarchy in the following diagram:

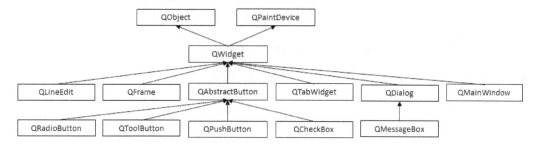

Figure 3.1 – QWidget class hierarchy

Most of the Qt widget names are self-explanatory and can be identified easily as they start with Q. Some of them are listed here:

- QPushButton is used to command an application to perform a certain action.

- QCheckBox allows the user to make a binary choice.

- QRadioButton allows the user to make only one choice from a set of mutually exclusive options.

- QFrame displays a frame.

- QLabel is used to display text or an image.

- QLineEdit allows the user to enter and edit a single line of plain text.

- QTabWidget is used to display pages related to each tab in a stack of tabbed widgets.

One of the advantages of using Qt Widgets is its parenting system. Any object that inherits from QObject has a parent-child relationship. This relationship makes many things convenient for developers, such as the following:

- When a widget is destroyed, all its children are destroyed as well due to the parent-children hierarchy. This avoids memory leaks.

- You can find children of a given QWidget class by using findChild() and findChildren().

- Child widgets in a Qwidget automatically appear inside the parent widget.

A typical C++ program terminates when the main returns, but in a GUI application we can't do that, or the application will be unusable. Thus, we will need the GUI to be present until the user closes the window. To accomplish this, the program should run in a loop till this happens. The GUI application waits for user input events.

Let's use QLabel to display a text with a simple GUI program, as follows:

```cpp
#include <QApplication>
#include <QLabel>
int main(int argc, char *argv[])
{
    QApplication app(argc, argv);
    QLabel myLabel;
    myLabel.setText("Hello World!");
    myLabel.show();
    return app.exec();
}
```

Remember to add the following line to the helloworld.pro file to enable the Qt Widgets module:

```
QT += widgets
```

You need to run qmake after you make changes to your .pro file. If you are using the command line, then proceed with the following commands:

```
>qmake
>make
```

Now, hit the **Run** button to build and run the application. You will soon see a UI with **Hello World!** displayed, as illustrated in the following screenshot:

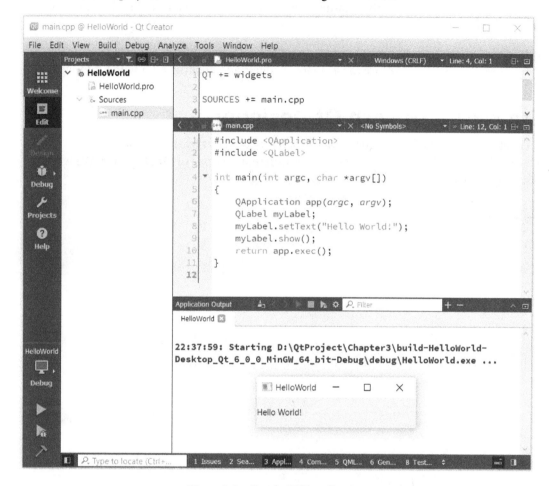

Figure 3.2 – Simple GUI application

You can also run the application from the command line on Windows, as follows:

```
>helloworld.exe
```

You can run the application from the command line on Linux distributions, as follows:

```
$./helloworld
```

In command-line mode, you may see a few error dialogs if the libraries are not found in the application path. You can copy the Qt libraries and plugin files to that binary folder to resolve the issue. To avoid these issues, we will stick to Qt Creator to build and run our sample programs.

In this section, we learned how to create a simple GUI using the Qt Widgets module. In the next section, we will explore the available widgets and creating a UI with Qt Designer.

Creating a UI with Qt Designer

Let's get familiar with Qt Designer's interface before we start learning how to design our own UI. The following screenshot shows different sections of **Qt Designer**. We will gradually learn about these sections while designing our UI:

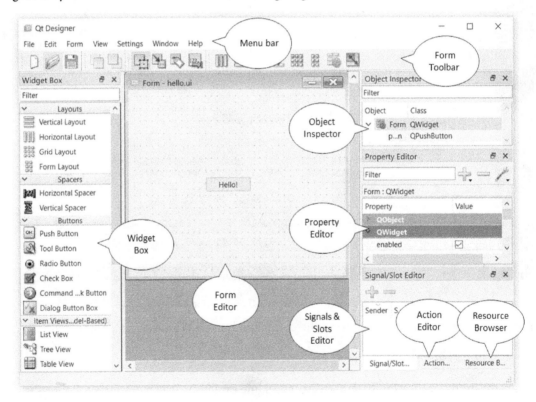

Figure 3.3 – Qt Designer UI

The Qt Widgets module comes with ready-to-use widgets. All these widgets can be found under the **Widget Box** section. Qt provides an option to create a UI by a drag-and-drop method. Let's explore these widgets by simply dragging them from the **Widget Box** area and dropping them into the **Form Editor** area. You can do this by grabbing an item and then pressing and releasing the mouse or trackpad over the intended region. Don't release your mouse or trackpad until the item reaches the **Form Editor** area.

The following screenshot shows different types of widgets available in the **Widget Box** section. We have added several ready-made widgets such as **Label**, **Push Button**, **Radio Button**, **Check Box**, **Combo Box**, **Progress Bar**, and **Line Edit** into the **Form Editor** area. These widgets are very commonly used widgets. You can explore the widget-specific properties in **Property Editor**:

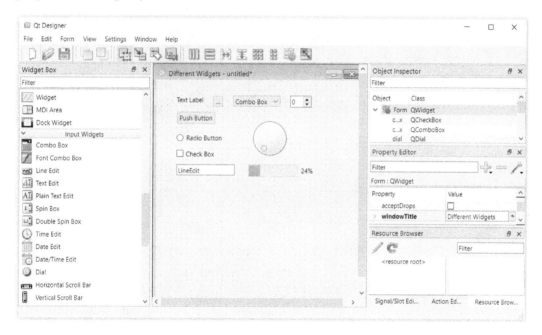

Figure 3.4 – Different types of GUI widgets

You can preview your UI by selecting the **Preview...** option under the **Form** menu, as shown in the following screenshot, or you can hit *Ctrl + R*. You will see a window with the UI preview:

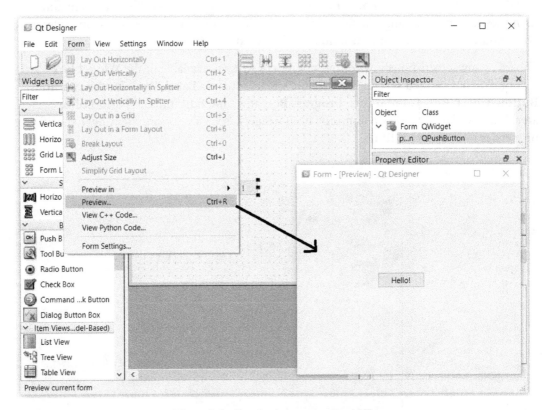

Figure 3.5 – Previewing your custom UI

You can find the created C++ code for your UI by selecting the **View C++ Code...** option under the **Form** menu, as shown in the following screenshot. You will see a window with the generated code. You can reuse the code while creating a dynamic UI:

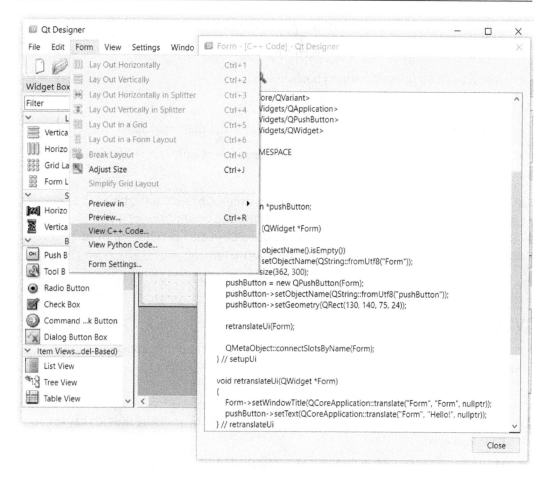

Figure 3.6 – Option to view corresponding C++ code

In this section, we got familiar with the Qt Designer UI. You can also find the same interface embedded in **Qt Creator** when you open a `.ui` file. In the next section, you will learn about different types of layouts and how to use them.

Managing layouts

Qt provides a set of convenient layout management classes to automatically arrange child widgets within another widget to ensure that the UI remains usable. The `QLayout` class is the base class of all layout managers. You can also create your own layout manager by reimplementing the `setGeometry()`, `sizeHint()`, `addItem()`, `itemAt()`, `takeAt()`, and `minimumSize()` functions. Please note that once the layout manager is deleted, the layout management will also stop.

The following list provides a brief description of the major layout classes:

- QVBoxLayout lines up widgets vertically.
- QHBoxLayout lines up widgets horizontally.
- QGridLayout lays widgets out in a grid.
- QFormLayout manages forms of input widgets and their associated labels.
- QStackedLayout provides a stack of widgets where only one widget is visible at a time.

QLayout uses multiple inheritances by inheriting from QObject and QLayoutItem. The subclasses of QLayout are QBoxLayout, QGridLayout, QFormLayout, and QStackedLayout. QVBoxLayout and QHBoxLayout are inherited from QBoxLayout with the addition of orientation information.

Let's use the Qt Designer module to lay out a few QPushButtons.

QVBoxLayout

In the QVBoxLayout class, widgets are arranged vertically, and they are aligned in the layout from top to bottom. At this point, you can do the following:

1. Drag four push buttons onto the **Form Editor**.

2. Rename the push buttons and select the push buttons by pressing the *Ctrl* key on your keyboard.

3. In the **Form** toolbar, click on the vertical layout button. You can find this by hovering on the toolbar button that says **Lay Out Vertically**.

    You can see the push buttons get arranged vertically in a top-down manner in the following screenshot:

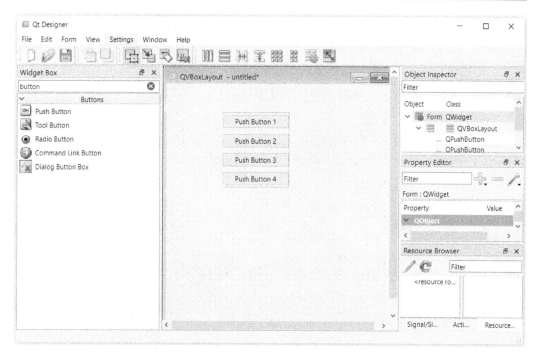

Figure 3.7 – Layout management with QVBoxLayout

You can also dynamically add a vertical layout through C++ code, as shown in the following snippet:

```
QWidget *widget = new QWidget;
QPushButton *pushBtn1 = new QPushButton("Push Button
                                    1");
QPushButton *pushBtn2 = new QPushButton("Push Button
                                    2");
QPushButton *pushBtn3 = new QPushButton("Push Button
                                    3");
QPushButton *pushBtn4 = new QPushButton("Push Button
                                    4");
QVBoxLayout *verticalLayout = new QVBoxLayout(widget);
verticalLayout->addWidget(pushBtn1);
verticalLayout->addWidget(pushBtn2);
verticalLayout->addWidget(pushBtn3);
verticalLayout->addWidget(pushBtn4);
widget->show ();
```

This program illustrates how to use a vertical layout object. Note that the `QWidget` instance, `widget`, will become the main window of the application. Here, the layout is set directly as the top-level layout. The first push button added to the `addWidget()` method occupies the top of the layout, while the last push button occupies the bottom of the layout. The `addWidget()` method adds a widget to the end of the layout, with a stretch factor and alignment.

If you don't set the parent window in the constructor, then you will have to use `QWidget::setLayout()` later to install the layout and reparent to the `widget` instance.

Next, we will look at the `QHBoxLayout` class.

QHBoxLayout

In the `QHBoxLayout` class, widgets are arranged horizontally, and they are aligned from left to right.

We can now do the following:

1. Drag four push buttons onto the **Form Editor**.

2. Rename the push buttons and select the push buttons by pressing the *Ctrl* key on your keyboard.

3. In the **Form** toolbar, click on the horizontal layout button. You can find this by hovering on the toolbar button that says **Lay Out Horizontally**.

   You can see the push buttons get arranged horizontally in a left-to-right manner in this screenshot:

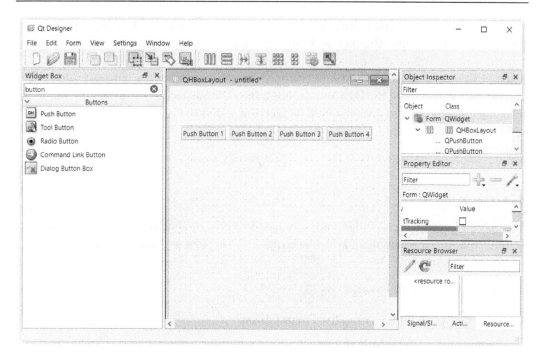

Figure 3.8 – Layout management with QHBoxLayout

You can also dynamically add a horizontal layout through C++ code, as shown in the following snippet:

```
QWidget *widget = new QWidget;
QPushButton *pushBtn1 = new QPushButton("Push
                                         Button 1");
QPushButton *pushBtn2 = new QPushButton("Push
                                         Button 2");
QPushButton *pushBtn3 = new QPushButton("Push
                                         Button 3");
QPushButton *pushBtn4 = new QPushButton("Push
                                         Button 4");
QHBoxLayout *horizontalLayout = new QHBoxLayout(
                                         widget);
horizontalLayout->addWidget(pushBtn1);
horizontalLayout->addWidget(pushBtn2);
horizontalLayout->addWidget(pushBtn3);
horizontalLayout->addWidget(pushBtn4);
widget->show ();
```

The preceding example demonstrates how to use a horizontal layout object. Similar to the vertical layout example, the QWidget instance will become the main window of the application. In this case, the layout is set directly as the top-level layout. By default, the first push button added to the addWidget() method occupies the leftmost side of the layout, while the last push button occupies the rightmost side of the layout. You can change the direction of growth when widgets are added to the layout by using the setDirection() method.

In the next section, we will look at the QGridLayout class.

QGridLayout

In the QGridLayout class, widgets are arranged in a grid by specifying the number of rows and columns. It resembles a grid-like structure with rows and columns, and widgets are inserted as items.

Here, we should do the following:

1. Drag four push buttons onto the **Form Editor**.
2. Rename the push buttons and select the push buttons by pressing the *Ctrl* key on your keyboard.
3. In the **Form** toolbar, click on the grid layout button. You can find this by hovering on the toolbar button that says **Lay Out in a Grid**.

    You can see the push buttons get arranged in a grid in the following screenshot:

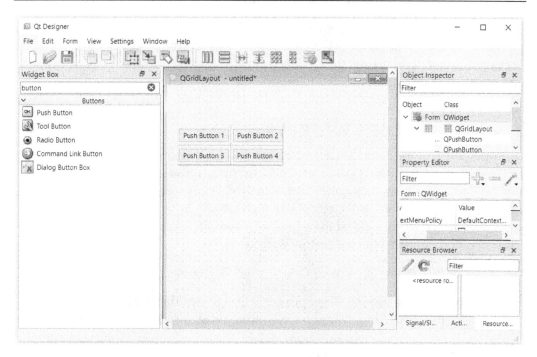

Figure 3.9 – Layout management with QGridLayout

You can also dynamically add grid layout through C++ code, as shown in the following snippet:

```
QWidget *widget = new QWidget;
QPushButton *pushBtn1 = new QPushButton(
                        "Push Button 1");
QPushButton *pushBtn2 = new QPushButton(
                        "Push Button 2");
QPushButton *pushBtn3 = new QPushButton(
                        "Push Button 3");
QPushButton *pushBtn4 = new QPushButton(
                        "Push Button 4");
QGridLayout *gridLayout = new QGridLayout(widget);
gridLayout->addWidget(pushBtn1);
gridLayout->addWidget(pushBtn2);
gridLayout->addWidget(pushBtn3);
gridLayout->addWidget(pushBtn4);
widget->show();
```

The preceding snippet explains how to use a grid layout object. The layout concept remains the same as in the previous sections. You can explore QFormLayout and QStackedLayout layouts from the Qt documentation. Let's proceed to the next section on how to create your custom widget and export it to the Qt Designer module.

Creating custom widgets

Qt provides ready-to-use essential **GUI elements**. Qt widgets were not actively developed after **Qt Modeling Language** (**QML**) came into existence, so you may require a more specific widget and may want to make it available to others. A custom widget may be a combination of one or more Qt widgets placed together or may be written from scratch. We will create a simple label widget from QLabel as our first custom widget. A custom widget collection can have multiple custom widgets.

Follow these steps to build your first Qt custom widgets library:

1. To create a new Qt custom widget project in Qt, click on the **File menu** option on the menu bar or hit *Ctrl + N*. Alternatively, you can also click on the **New Project** button located on the **Welcome** screen. Select the **Other Project** template and then select **Qt Custom Designer Widget**, as shown in the following screenshot:

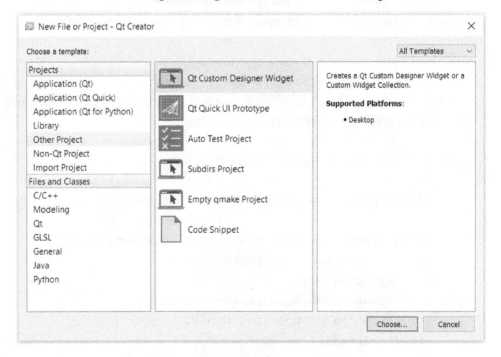

Figure 3.10 – Creating a custom widget library project

2. In the next step, you will be asked to choose the project name and project location. You can navigate to the desired project location by clicking the **Browse…** button. Let's name our sample project MyWidgets. Then, click on the **Next** button to proceed to the next screen. The following screenshot illustrates this step:

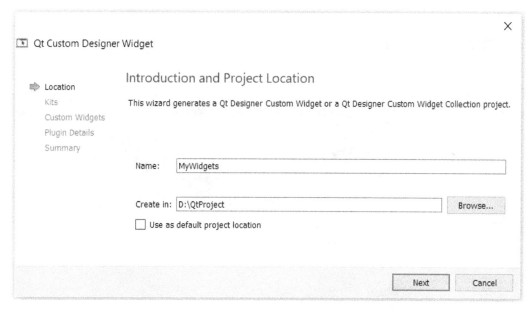

Figure 3.11 – Creating custom controls library project

3. In the next step, you can select a kit from a set of kits to build and run your project. To build and run the project, at least one kit must be active and selectable. Select the default **Desktop Qt 6.0.0 MinGW 64-bit** kit. Click on the **Next** button to proceed to the next screen. The following screenshot illustrates this step:

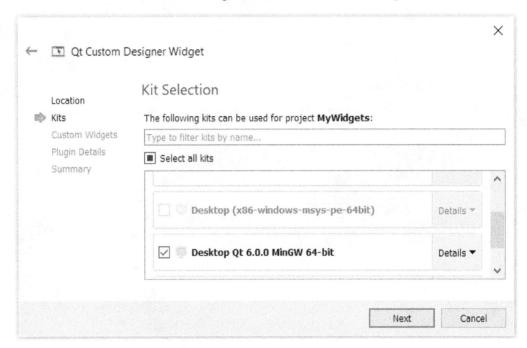

Figure 3.12 – Kit selection screen

4. In this step, you can define your custom widget class name and inheritance details. Let's create our own custom label with the class name `MyLabel`. Click on the **Next** button to proceed to the next screen. The following screenshot illustrates this step:

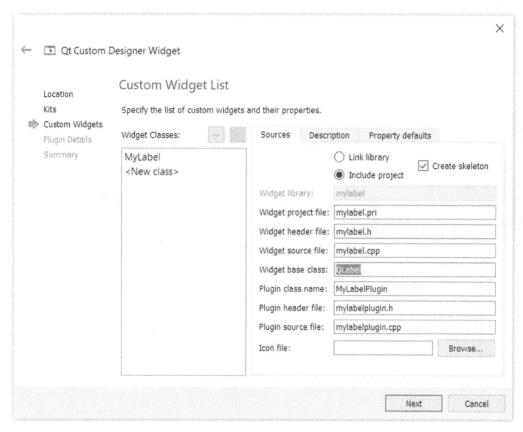

Figure 3.13 – Creating a custom widget from an existing widget's screen

5. In the next step, you can add more custom widgets to create a widget collection. Let's create our own custom frame with the class name `MyFrame`. You can add more information to the **Description** tab or can modify it later. Click on the checkbox that says **The widget is a container** to use the frame as a container. Click on the **Next** button to proceed to the next screen. The following screenshot illustrates this step:

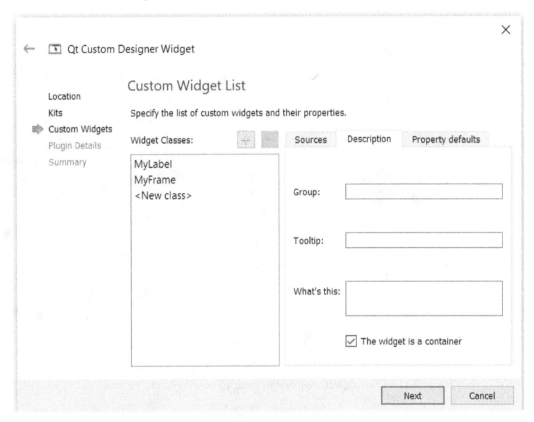

Figure 3.14 – Creating a custom widget container

6. In this step, you can specify the collection class name and the plugin information to automatically generate the project skeleton. Let's name the collection class `MyWidgetCollection`. Click on the **Next** button to proceed to the next screen. The following screenshot illustrates this step:

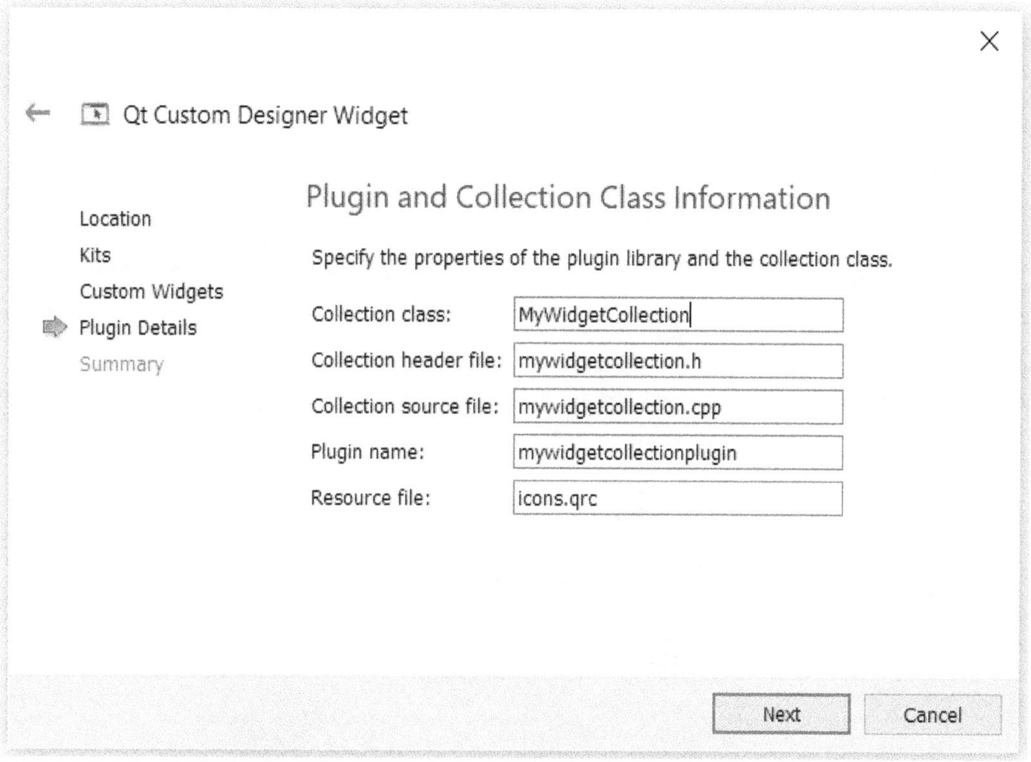

Figure 3.15 – Option to specify plugin and collection class information

7. The next step is to add your custom widget project to the installed version control system. You may skip version control for this project. Click on the **Finish** button to create the project with the generated files. The following screenshot illustrates this step:

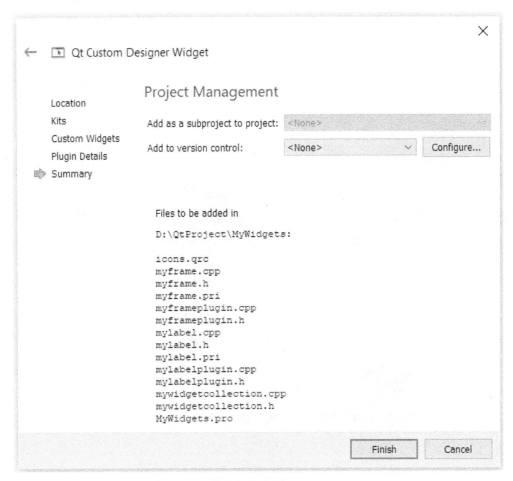

Figure 3.16 – Project management screen

8. Expand the **Project Explorer** views and open the mylabel.h file. We will modify the contents to extend the functionalities. Add a QDESIGNER_WIDGET_EXPORT macro before the custom widget class name to ensure the class is exported properly in the **dynamic-link library (DLL)** or the shared library. Your custom widget may work without this macro, but it is a good practice to add this macro. You will have to add #include <QtDesigner> to the header file after you insert the macro. The following screenshot illustrates this step:

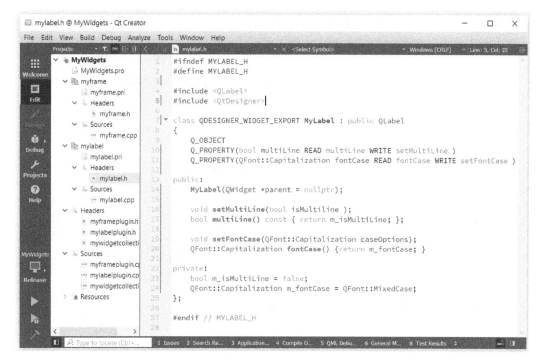

Figure 3.17 – Modifying the custom widget from the created skeleton

> **Important note**
>
> On some platforms, the build system may remove the symbols required by Qt Designer module to create new widgets, making them unusable. Using the QDESIGNER_WIDGET_EXPORT macro ensures that the symbols are retained on those platforms. This is important while creating a cross-platform library. There are no side effects on other platforms.

9. Now, open the `mylabelplugin.h` file. You will find that the plugin class is inherited from a new class named QDesignerCustomWidgetInterface. This class allows Qt Designer to access and create custom widgets. Please note that you must update the header file as follows to avoid deprecated warnings:

```
#include <QtUiPlugin/QDesignerCustomWidgetInterface>
```

10. You will find several functions auto created in `mylabelplugin.h`. Don't remove these functions. You can specify the values in the `name()`, `group()`, and `icon()` functions that appear in the Qt Designer module. Note that if you don't specify an icon path in `icon()`, then Qt Designer will use the default Qt icon. The `group()` function is illustrated in the following code snippet:

```
QString MyFramePlugin::group() const
{
    return QLatin1String("My Containers");
}
```

11. You can see in the following code snippet that `isContainer()` returns `false` in `MyLabel` and `true` in `MyFrame`, since `MyLabel` is not designed to hold other widgets. Qt Designer calls `createWidget()` to obtain an instance of `MyLabel` or `MyFrame`:

```
bool MyFramePlugin::isContainer() const
{
    return true;
}
```

12. To create a widget with a defined geometry or any other properties, you specify these in the `domXML()` method. The function returns an **Extensible Markup Language (XML)** snippet that is used by the widget factory to create a custom widget with the defined properties. Let's specify the `MyLabel` width as `100` **pixels (px)** and height as `16` px, as follows:

```
QString MyLabelPlugin::domXml() const
{
    return "<ui language=\"c++\""
              displayname=\"MyLabel\">\n"
           " <widget class=\"MyLabel\""
              name=\"myLabel\">\n"
           "  <property name=\"geometry\">\n"
           "   <rect>\n"
           "    <x>0</x>\n"
           "    <y>0</y>\n"
           "    <width>100</width>\n"
           "    <height>16</height>\n"
```

```
    "    </rect>\n"
    "   </property>\n"
    "   <property name=\"text\">\n"
    "    <string>MyLabel</string>\n"
    "   </property>\n"
    "  </widget>\n"
    "</ui>\n";
}
```

13. Now, let's have a look at the `MyWidgets.pro` file. It contains all the information required by qmake to build the custom widget collection library. You can see in the following code snippet that the project is a library type and is configured to be used as a plugin:

```
CONFIG          += plugin debug_and_release
CONFIG          += c++17
TARGET          = $$qtLibraryTarget(
                  mywidgetcollectionplugin)
TEMPLATE        = lib
HEADERS         = mylabelplugin.h myframeplugin.h
mywidgetcollection.h
SOURCES         = mylabelplugin.cpp myframeplugin.cpp \
                          mywidgetcollection.cpp
RESOURCES       = icons.qrc
LIBS            += -L.
greaterThan(QT_MAJOR_VERSION, 4) {
    QT += designer
} else {
    CONFIG += designer
}
target.path = $$[QT_INSTALL_PLUGINS]/designer
INSTALLS        += target
include(mylabel.pri)
include(myframe.pri)
```

14. We have gone through the custom widget creation process. Let's run qmake and build the library in the **Release** mode. Right-click on the project and click on the **Build** option, as shown in the following screenshot. The project will be built within a few seconds and will be available inside the `inside release` folder. On the Windows platform, you can manually copy the `mywidgetcollectionplugin.dll` created plugin library to the `D:\Qt\6.0.0\mingw81_64\plugins\designer` path. This path and extension vary for different operating systems:

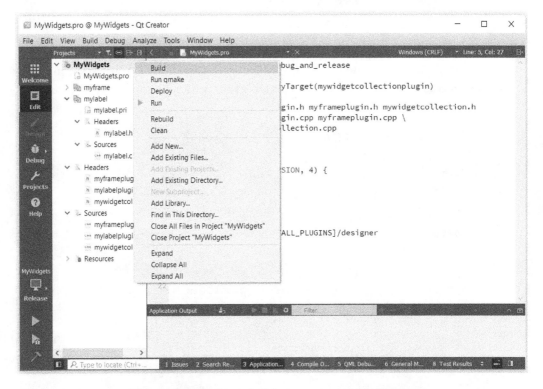

Figure 3.18 – Option to build your custom widget library

15. We have created our custom plugin. Now, close the plugin project and click on the designer.exe file present inside D:\Qt\6.0.0\mingw81_64\bin. You can see MyFrame under the **Custom Widgets** section, as shown in the following screenshot. Click on the **Create** button or use a widget template. You can also register your own form as a template by doing platform-specific modifications. Let's use the Qt Designer-provided widget template:

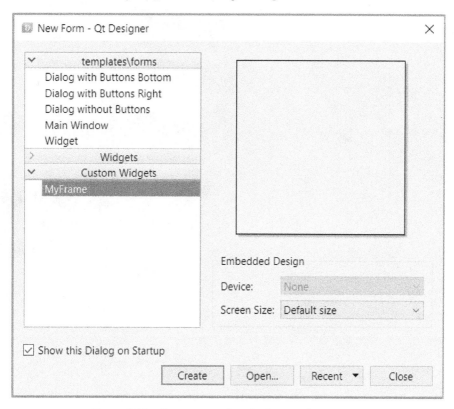

Figure 3.19 – Custom container in the new form screen

16. You can see our custom widgets in the left-side **Widget Box** section, at the bottom. Drag the **MyLabel** widget to the form. You can find created properties such as **multiLine** and **fontCase** along with **QLabel** properties under **Property Editor**, as illustrated in the following screenshot:

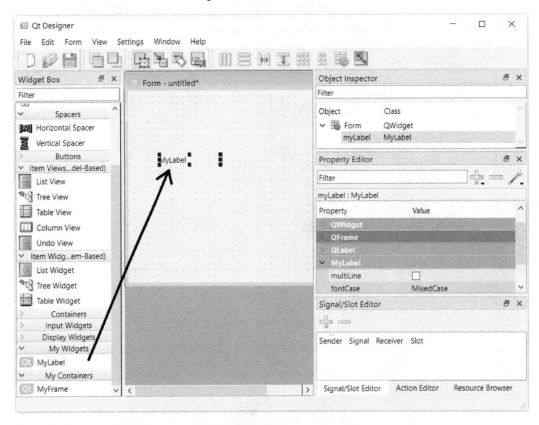

Figure 3.20 – Exported widgets available in Qt Designer

You can also find detailed instructions with examples in the following Qt documentation link:

`https://doc.qt.io/qt-6/designer-creating-custom-widgets.html`

Congratulations! You have successfully created your custom widgets with new properties. You can create complex custom widgets by combining multiple widgets. In the next section, you will learn how to customize the look and feel of widgets.

Creating Qt Style Sheets and custom themes

In the last section, we created our custom widget, but the widget still has a native look. Qt provides several ways to customize the look and feel of the UI. A **Qt Style Sheet** is one of the simplest ways to change the look and feel of widgets without doing much complex coding. Qt Style Sheet syntax is identical to **HyperText Markup Language/Cascading Style Sheets (HTML/CSS)** syntax. Style Sheets comprise a sequence of style rules. A style rule consists of a selector and a declaration. The selector specifies widgets that will be affected by the style rule, and the declaration specifies the properties of the widget. The declaration portion of a style rule is a list of properties as key-value pairs, enclosed inside { } and separated by semicolons.

Let's have look at the simple QPushButton Style Sheet syntax, as follows:

```
QPushButton { color: green; background-color: rgb
(193, 255, 216);}
```

You can also change the look and feel of widgets by applying Style Sheet in Qt Designer with the stylesheet editor, as follows:

1. Open the Qt Designer module and create a new form. Drag and place a push button on the form.

2. Then, right-click on the push button or anywhere in the form to get the context menu.

3.   Next, click on the **Change styleSheet…** option, as shown in the following screenshot:

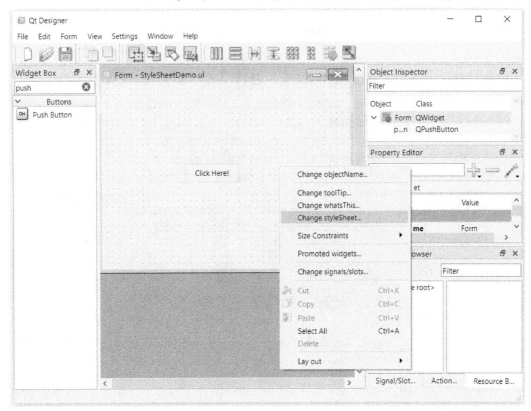

Figure 3.21 – Adding Style Sheet using Qt Designer

4.   We have used the following Style sheet to create the previous look and feel. You can also change a Style Sheet from the `QWidget` properties in **Property Editor**:

```
QPushButton {
    background-color: rgb(193, 255, 216);
    border-width: 2px;
    border-radius: 6;
    border-color: lime;
    border-style: solid;
    padding: 2px;
    min-height: 2.5ex;
    min-width: 10ex;
}
QPushButton:hover {
```

```
        background-color: rgb(170, 255, 127);
    }
QPushButton:pressed {
        background-color: rgb(170, 255, 127);
        font: bold;
    }
```

In the preceding example, only `Push Button` will get the style described in the Style Sheet, and all other widgets will have the native styling. You can also create different styles for each push button and apply the styles to respective push buttons by mentioning their object names in the Style Sheet, as follows:

```
QPushButton#pushButtonID
```

> **Important note**
>
> To learn more about Style Sheet and their usage, read the documentation at the following links:
>
> https://doc.qt.io/qt-6/stylesheet-reference.html
>
> https://doc.qt.io/qt-6/stylesheet-syntax.html
>
> https://doc.qt.io/qt-6/stylesheet-customizing.html

Using a QSS file

You can combine all Style Sheet code in a defined `.qss` file. This helps in ensuring the look and feel is maintained across the application in all screens. QSS files are analogous to `.css` files, which contain the definitions for the look and feel of GUI elements such as color, background color, font, and mouse interaction behaviors. They can be created and edited with any text editor. You can create a new Style Sheet file with the `.qss` file extension and then add it to the resource file (`.qrc`). You may or may not have `.ui` files for all projects. The GUI controls can be created dynamically through code. You can apply a Style Sheet to a widget or to a whole application, as shown in the following code snippet. This is how we do it for a custom widget or form:

```
MyWidget::MyWidget(QWidget *parent)
    : QWidget(parent)
{
    setStyleSheet("QWidget { background-color: green }");
}
```

Here is how we apply it for the whole application:

```
#include "mywidget.h"
#include <QApplication>
#include <QFile>
int main(int argc, char *argv[])
{
    QApplication app(argc, argv);
    QFile file(":/qss/default.qss");
    file.open(QFile::ReadOnly);
    QString styleSheet = QLatin1String(file.readAll());
    app.setStyleSheet(styleSheet);
    Widget mywidget;
    mywidget.show();
    return app.exec();
}
```

The preceding program illustrates how to use a Style Sheet file for the entire Qt GUI application. You need to add the .qss file into the resources. Open the .qss file using QFile and pass the customized QSS rules as an argument to the setStyleSheet() method on the QApplication object. You will see all screens will have the Style Sheet applied to them.

In this section, you learned about customizing the look and feel of an application using Style Sheets, but there are more ways to change an application's look and feel. These approaches depend on your project need. In the next section, you will learn about custom styles.

Exploring custom styles

Qt provides several QStyle subclasses that emulate the styles of the different platforms supported by Qt. These styles are readily available with the Qt GUI module. You can build your own **custom styles** and export these as plugins. Qt uses QStyle for rendering the Qt widgets to ensure their look and feel, as per native widgets.

On a Unix distribution, you can get a Windows-style UI for your application by running the following command:

```
$ ./helloworld -style windows
```

You can set a style on an individual widget using the `QWidget::setStyle()` method.

Creating a custom style

You can customize the look and feel of your GUI by creating a custom style. There are two different approaches to creating a custom style. In a static approach, you can subclass the `QStyle` class and reimplement virtual functions to deliver the desired behavior, or rewrite the `QStyle` class from scratch. `QCommonStyle` is generally used as a base class instead of `QStyle`. In a dynamic approach, you can subclass `QProxyStyle` and modify the behavior of your system style at runtime. You can also develop style-aware custom widgets by using `QStyle` functions such as `drawPrimitive()`, `drawItemText()`, and `drawControl()`.

This section is an advanced Qt topic. You need to understand Qt in depth to create your own style plugin. You can skip this section if you are a beginner. You can learn about the QStyle classes and custom styles in the Qt documentation at the following link:

```
https://doc.qt.io/qt-6/qstyle.html
```

Using a custom style

There are several ways to apply a custom style in a Qt application. The easiest way is to call the `QApplication::setStyle()` static function before creating a `QApplication` object, as follows:

```cpp
#include "customstyle.h"
int main(int argc, char *argv[])
{
    QApplication::setStyle(new CustomStyle);
    QApplication app(argc, argv);
    Widget helloworld;
    helloworld.show();
    return app.exec();
}
```

You can also apply a custom style as a command-line argument, like so:

```
>./customstyledemo -style customstyle
```

Custom styles can be difficult to implement but can be faster and more flexible. QSS is easy to learn and implement, but the performance may get affected, especially at the application launch time, as the QSS parsing may take time. You can choose the approach convenient to you or your organization. We have learned how to customize the GUI. Now, let's understand what widgets, windows, and dialogs are in the last section of this chapter.

Using widgets, windows, and dialogs

A widget is a GUI element that can be displayed on the screen. This could include labels, push buttons, list views, windows, dialogs, and so on. All widgets display certain information to a user on the screen, and most of them allow user interactions through the keyboard or mouse.

A window is a top-level widget that doesn't have another parent widget. Generally, windows have a title bar and border unless any window flags are specified. The window style and certain policies are determined by the underlying windowing system. Some of the common window classes in Qt are QMainWindow, QMessageBox, and QDialog. A main window usually follows a predefined layout for desktop applications that comprises a menu bar, a toolbar, a central widget area, and a status bar. QMainWindow requires a central widget even if it is just a placeholder. Other components can be removed in a main window. *Figure 3.22* illustrates the layout structure of QMainWindow. We typically call the show() method to display a widget or main window.

QMenuBar is present at the top of QMainWindow. You can add menu options such as **File**, **Edit**, **View**, and **Help**. In the following screenshot showing QMenuBar, there is QToolBar. QDockWidget provides a widget that can be docked inside QMainWindow or floated as a top-level window. The central widget is the primary view area where you can add your form or child widgets. Create your own view area with child widgets and then call setCentralWidget():

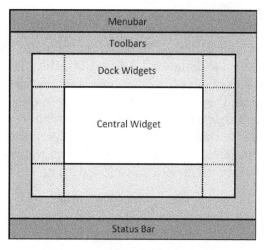

Figure 3.22 – QMainWindow layout

> **Important note**
>
> QMainWindow shouldn't be confused with QWindow. QWindow is
> a convenient class that represents a window in the underlying windowing
> system. Usually, applications use QWidget or QMainWindow for their UI.
> However, it is possible to render directly to QWindow, if you want to keep
> minimal dependencies.

Dialogs are temporary windows that are used to provide notifications or receive user inputs and usually have **OK** and **Cancel**-type buttons. QMessageBox is a type of dialog that is used to show information and alerts or to ask a question to the user. Typically, the exec() method is used to show a dialog. The dialog is shown as a modal dialog and is blocking in nature until the user closes it. A simple message box can be created with the following code snippet:

```
QMessageBox messageBox;
messageBox.setText("This is a simple QMessageBox.");
messageBox.exec();
```

The takeaway is that all of these are widgets. Windows are the top-level widgets, and dialogs are a special kind of window.

Summary

This chapter explained the fundamentals of the Qt Widgets module and how to create a custom UI. Here, you learned to design and build GUIs with Qt Designer. Traditional desktop applications are usually built with Qt Designer. Features such as custom widget plugins allow you to create and use your own widget collection with Qt Designer. We also discussed customizing the look and feel of your application with style sheets and styles, as well as looking at the uses of and differences between widgets, windows, and dialogs. Now, you can create a GUI application with your own custom widgets with extended functionalities and create your own themes for your desktop application.

In the next chapter, we will discuss QtQuick and QML. Here, you will learn about QtQuick controls, Qt Quick Designer, and how to build a custom QML application. We will also discuss an alternate option of using Qt Quick for GUI design rather than widgets.

4
Qt Quick and QML

Qt consists of two different modules for developing a **graphical user interface** (**GUI**) application. The first approach is to use Qt Widgets and C++, which we learned about in the previous chapter. The second approach is to use Qt Quick Controls and the **Qt Modeling Language** (**QML**), which we will be covering in this chapter.

In this chapter, you will learn how to use Qt Quick Controls and the QML scripting language. You will study how to use Qt Quick Layouts and positioners and make a responsive GUI application. You will learn to integrate your backend C++ code with frontend QML. You will learn the fundamentals of Qt Quick and QML, and how to develop touch-friendly and visual-oriented Qt applications. You will also learn about mouse and touch events and how to develop a touch-aware application.

In this chapter, we're going to cover the following main topics:

- Getting started with QML and Qt Quick
- Understanding Qt Quick Controls
- Creating a simple Qt Quick application
- Designing a **user interface** (**UI**) with Qt Quick Designer
- Positioners and layouts in QML
- Integrating QML with C++

- Integrating QML with **JavaScript (JS)**
- Handling mouse and touch events

By the end of this chapter, you will understand the basics of QML, integration with C++, and how to create your own fluid UI.

Technical requirements

The technical requirements for this chapter include minimum versions of Qt 6.0.0 and Qt Creator 4.14.0 installed on the latest desktop platforms such as Windows 10, Ubuntu 20.04, or macOS 10.14.

All the code used in this chapter can be downloaded from the following GitHub link: `https://github.com/PacktPublishing/Cross-Platform-Development-with-Qt-6-and-Modern-Cpp/tree/master/Chapter04`.

> **Important note**
> The screenshots used in this chapter are taken from the Windows platform. You will see similar screens based on the underlying platforms in your machine.

Getting started with QML and Qt Quick

QML is a UI markup language. It is a declarative language that is part of the Qt framework. It enables the building of fluid and touch-friendly UIs and came into existence with the evolution of touchscreen mobile devices. It was created to be highly dynamic, where developers can easily create fluid UIs with minimal coding. The Qt QML module implements the QML architecture and provides a framework for developing applications. It defines and implements the language and infrastructure, and provides **application programming interfaces (APIs)** to integrate the QML language with JS and C++.

Qt Quick provides a library of types and functionality for QML. It comprises interactive types, visual types, animations, models, views, and graphics effects. It is used for mobile applications where touch input, fluid animations, and user experience are crucial. The Qt QML module provides the language and infrastructure for QML applications, whereas the Qt Quick module provides many visual elements, animation, and many more modules to develop touch-oriented and visually appealing applications. Instead of using Qt Widgets for UI design, you can use QML and Qt Quick Controls. Qt Quick supports several platforms, such as Windows, Linux, Mac, iOS, and Android. You can create a custom class in C++ and port it over to Qt Quick to extend its functionality. Furthermore, the language provides a smooth integration with C++ and JS.

Understanding the QML type system

Let's get familiar with the **QML type system** and various QML types. The types in a QML file can originate from various sources. The different types used in a QML file are outlined here:

- Basic types provided natively by QML such as `int`, `bool`, `real`, and `list`
- JS types such as `var`, `Date`, and `Array`
- QML object types such as `Item`, `Rectangle`, `Image`, and `Component`
- Types registered via C++ by QML modules such as `BackendLogic`
- Types provided as a QML file, such as `MyPushButton`

A basic type can contain a simple value such as an `int` or a `bool` type. In addition to the native basic types, the Qt Quick module also provides additional basic types. The QML engine also supports JS objects and arrays. Any standard JS type can be created and stored using the generic `var` type. Please note that the `variant` type is obsolete and exists only to support older applications. A QML object type is a type from which a QML object can be created. Custom QML object types can be defined by creating a `.qml` file that defines the type. QML object types can have properties, methods, signals, and so on.

To use the basic QML types inside your QML file, import the `QtQml` module with the following line of code: `import QtQml`

`Item` is the base type for all visual elements in Qt Quick. All visual items in Qt Quick are inherited from `Item`, which is a transparent visual element that can be used as a container. Qt Quick provides `Rectangle` as a visual type to draw rectangles, and an `Image` type to display images. `Item` provides a common set of properties for the visual elements. We will explore the usage of these types throughout the book.

You can learn more about QML types at the following link:

`https://doc.qt.io/qt-6/qmltypes.html`

In this section, we learned the basics of QML and Qt Quick. In the next section, we will discuss Qt Quick Controls.

Understanding Qt Quick Controls

Qt Quick Controls provides a set of UI elements that can be used to build a fluid UI using Qt Quick. To avoid ambiguity with **widgets**, we will use the term **controls** for UI elements. **Qt Quick Controls 1** was originally designed to support desktop platforms. With the evolution of mobile devices and embedded systems, the module required changes to meet performance expectations. Hence, **Qt Quick Controls 2** was born, and it further enhanced support for mobile platforms. Qt Quick Controls 1 has been deprecated since Qt 5.11 and has been removed from Qt 6.0. Qt Quick Controls 2 is now simply known as Qt Quick Controls.

The QML types can be imported into your application using the following `import` statement in your `.qml` file:

```
import QtQuick.Controls
```

> **Important note**
>
> In Qt 6, there are certain changes in the QML import and versioning system. The version numbers have been kept optional. If you import a module without specifying the version number, then the latest version of the module is imported automatically. If you import a module with only the major version number, then the module is imported with a specified major version and the latest minor version. Qt 6 introduced an **auto imports** functionality, which is written as `import <module> auto`. This ensures the imported module and importing module have the same version number.
>
> Changes to Qt Quick Controls in Qt 6 can be found at the following link:
>
> `https://doc.qt.io/qt-6/qtquickcontrols-changes-qt6.html`

Qt Quick Controls offers QML types for creating UIs. Example of Qt Quick Controls are given here:

- `ApplicationWindow`: Styled top-level window with support for a header and footer
- `BusyIndicator`: Indicates background activity—for instance, while content is being loaded
- `Button`: Push button that can be clicked to perform a command or answer a question
- `CheckBox`: Check button that can be toggled on or off
- `ComboBox`: Combined button and pop-up list for selecting options

- `Dial`: Circular dial that is rotated to set a value

- `Dialog`: Pop-up dialog with standard buttons and a title

- `Label`: Styled text label with inherited font

- `Popup`: Base type of pop-up-like UI controls

- `ProgressBar`: Indicates the progress of an operation

- `RadioButton`: Exclusive radio button that can be toggled on or off

- `ScrollBar`: Vertical or horizontal interactive scroll bar

- `ScrollView`: Scrollable view

- `Slider`: Used to select a value by sliding a handle along a track

- `SpinBox`: Allows the user to select from a set of preset values

- `Switch`: Button that can be toggled on or off

- `TextArea`: Multiline text-input area

- `TextField`: Single-line text input field

- `ToolTip`: Provides tool tips for any control

- `Tumbler`: Spinnable wheel of items that can be selected

To configure the Qt Quick Controls module for building with qmake, add the following line to the project's `.pro` file:

```
QT += quickcontrols2
```

In this section, we learned about the different types of UI elements available with Qt Quick. In the next section, we will discuss the different styles provided by Qt Quick and how to apply them.

Styling Qt Quick Controls

Qt Quick Controls comes with a standard set of styles. They are listed here:

- **Basic**

- **Fusion**

- **Imagine**

- **Material**

- **Universal**

There are two ways to apply styles in Qt Quick Controls, as follows:

- Compile time
- Runtime

You can apply a compile-time style by importing the corresponding style module, as shown here:

```
import QtQuick.Controls.Universal
```

You can apply a runtime style by using one of the following approaches:

QQuickStyle::setStyle()	`QQuickStyle::setStyle("Universal");`
-style command line argument	`./application -style universal`
Environment variable QT_QUICK_CONTROLS_STYLE	`QT_QUICK_CONTROLS_STYLE= universal ./application`
qtquickcontrols2.conf configuration file	`[Controls] Style=Universal`

Figure 4.1 – Different ways to apply a style at runtime

In this section, we learned about the available styles in Qt Quick. In the next section, we will create our first Qt Quick GUI application.

Creating a simple Qt Quick application

Let's create our first Qt Quick application using Qt 6. A Hello World program is a very simple program that displays Hello World!. The project uses minimal—and the most basic—code. For this project, we will use the **project skeleton** created by Qt Creator. So, let's begin! Proceed as follows:

1. To create a new Qt Quick application, click on the **File menu** option on the menu bar or hit *Ctrl + N*. Alternatively, you can also click on the **New Project** button located on the welcome screen. Then, a window will pop up for you to choose a project template. Select **Qt Quick Application - Empty** and click the **Choose...** button, as shown in the following screenshot:

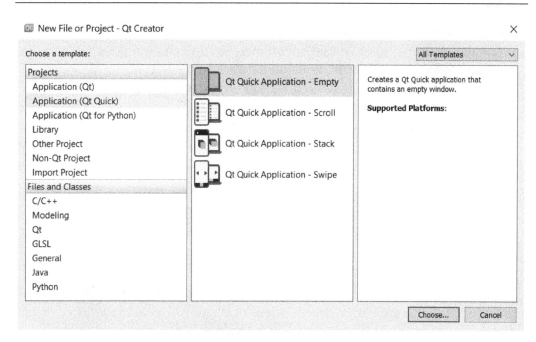

Figure 4.2 – New Qt Quick application wizard

2. In the next step, you will be asked to choose a project name and a project location. You can navigate to the desired project location by clicking the **Browse...** button. Let's name our sample project SimpleQtQuickApp. Then, click on the **Next** button to proceed to the next screen, as shown in the following screenshot:

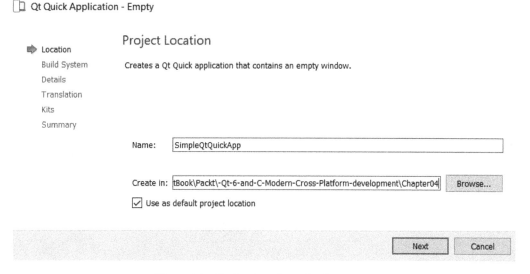

Figure 4.3 – Project location selection screen

3. In the next step, you can select a kit from a set of kits to build and run your project. To build and run a project, at least one kit must be active and selectable. Select the default **Desktop Qt 6.0.0 MinGW 64-bit** kit. Click on the **Next** button to proceed to the next screen. This can be seen in the following screenshot:

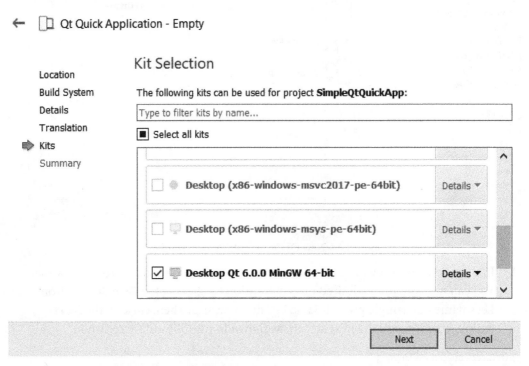

Figure 4.4 – Kit selection screen

4. The next step is to add your Qt Quick project to the installed **version control system** (**VCS**). You may skip version control for this project. Click on the **Finish** button to create a project with the generated files, as shown in the following screenshot:

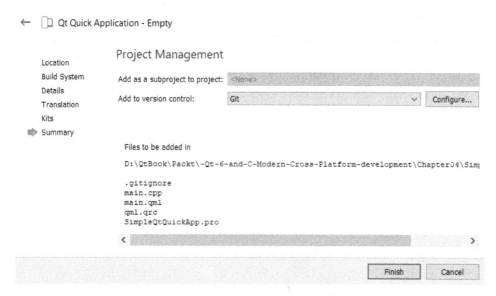

Figure 4.5 – Project management screen

5. Once a project has been created, Qt Creator will automatically open up a file from your project, called main.qml. You will see a type of script that is very different from your usual C/C++ projects, as shown in the following screenshot:

Figure 4.6 – Code editor screen showing the main.qml file

The QML runtime is implemented in C++ in the QtQml module. It contains a QML engine that is responsible for the execution of QML. It also holds the contexts and properties that will be accessible for the QML elements. Qt provides a QQmlEngine class for instantiating the QML components. You can also use the QQmlApplicationEngine class to load the application with a single QML file in a convenient way, as shown here:

```cpp
#include <QGuiApplication>
#include <QQmlApplicationEngine>
int main(int argc, char *argv[])
{
    QGuiApplication app(argc, argv);
    QQmlApplicationEngine engine;
    const QUrl url(QStringLiteral("qrc:/main.qml"));
    engine.load(url);
    return app.exec();
}
```

You can also use the QQuickView class, which provides a window for displaying a Qt Quick UI. This approach is little old. QQmlApplicationEngine has a convenient central application functionality with QML, whereas QQuickView is normally controlled from C++. The following code snippet shows how to use QQuickView to load a .qml file:

```cpp
#include <QGuiApplication>
#include <QQuickView>
int main(int argc, char *argv[])
{
    QGuiApplication app(argc, argv);
    QQuickView view;
    view.setResizeMode(
        QQuickView::SizeRootObjectToView);
    view.setSource(QUrl("qrc:/main.qml"));
    view.show();
    return app.exec();
}
```

QQuickView doesn't support using Window as a root item. If you want to create your root window from QML, then opt for QQmlApplicationEngine. While using QQuickView, you can directly use any Qt Quick element, as shown in the following code snippet:

```
import QtQuick
Item {
    width: 400
    height: 400
    Text {
        anchors.centerIn: parent
        text: "Hello World!"
    }
}
```

6. Next, you can build and run the Qt Quick project by clicking on the green arrow button located at the bottom-left corner of the **integrated development environment (IDE)**, as shown in the following screenshot:

Figure 4.7 – The build and run option in Qt Creator

7. Now, hit the **Run** button to build and run the application. Soon, you will see a UI with **Hello World!**, as shown in the following screenshot:

Figure 4.8 – Output of the Hello World UI

You can run the application from the command line on Windows, as follows:

```
>SimpleQtQuickApp.exe
```

You can also run the application from the command line on Linux distributions, as follows:

```
$./SimpleQtQuickApp
```

In command-line mode, you may see a few error dialogs if the libraries are not found in the application path. You can copy the Qt libraries and plugin files to that binary folder to resolve the issue. To avoid these issues, we will stick to Qt Creator to build and run our sample programs. You can switch between different kits by going to the project interface and selecting a kit based on your preferences. Please remember that you need to run qmake after you make changes to your .pro file. If you are using the command line, then proceed with the following commands:

```
>qmake
>make
```

You can also create a Qt Quick 2 UI project with a QML entry point without using any C++ code. To use it, you need to have a QML runtime environment such as `qmlscene` set up. Qt Creator uses `.qmlproject` to handle QML-only projects:

1. To create a Qt Quick 2 UI project, select **Qt Quick 2 UI Prototype** from the new project template screen, as shown in the following screenshot:

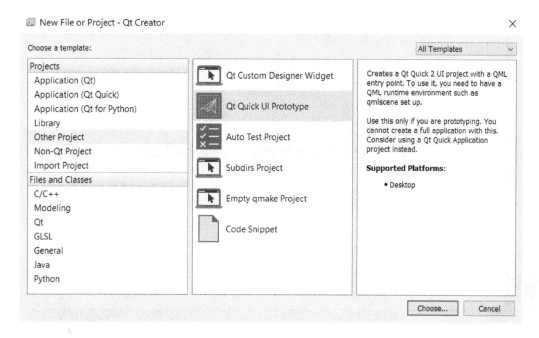

Figure 4.9 – Qt Quick UI Prototype wizard

2. Continue clicking the **Next** button to see the **Project Details**, **Kit Selection**, and **Project Management** screens. These screens are the same as for the Qt quick application project discussed earlier. Click on the **Finish** button to create a project with a skeleton. Now, have a look at the contents of the `QtQuickUIPrototype.qmlproject` and `QtQuickUIPrototype.qml` Qt Creator-generated files.

3. Let's modify the contents of `QtQuickUIPrototype.qml` to add a `Text` element and display `Hello World!`, as illustrated in the following screenshot:

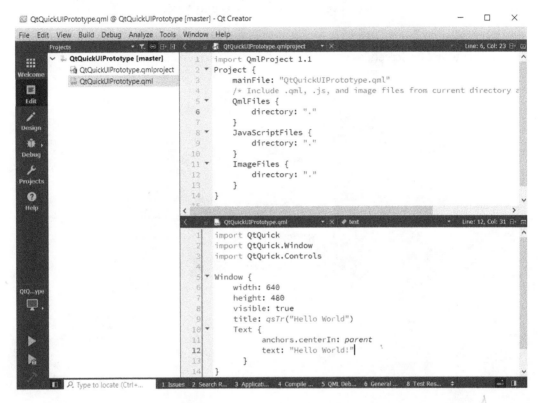

Figure 4.10 – Sample contents of Qt Quick UI Prototype project

4. Now, hit the **Run** button to build and run the application. Soon, you will see a UI with **Hello World!**.

You can also run the application from the command line, as follows:

```
>qmlscene QtQuickUIPrototype.qml
```

You may have to mention `qmlscene` and the `qml` file path in the command line. Use this only if you are prototyping. You cannot create a full application with this. Consider using a Qt Quick application project instead for a full application.

In this section, we learned how to create a simple GUI using the Qt Quick module. In the next section, we will learn how to design a custom UI using the Qt Quick Designer UI.

Designing a UI with Qt Quick Designer

In this section, you will learn how to use Qt Quick Designer to design your UI. Similar to the .ui file in Qt Widgets, you can also create a UI file in QML. The file has a .ui. qml file extension. There are two types of QML file: one with a .qml extension and another with a .ui.qml extension. The QML engine treats it as a standard .qml file, but it prohibits the logical implementation inside it. It creates a reusable UI definition for multiple .qml files. Through the separation of UI definition and logical implementation, it enhances the maintainability of QML code.

Let's get familiar with Qt Quick Designer's interface before we start learning how to design our own UI. The following screenshot shows different sections of Qt Quick Designer. We will gradually learn about these sections while designing our UI:

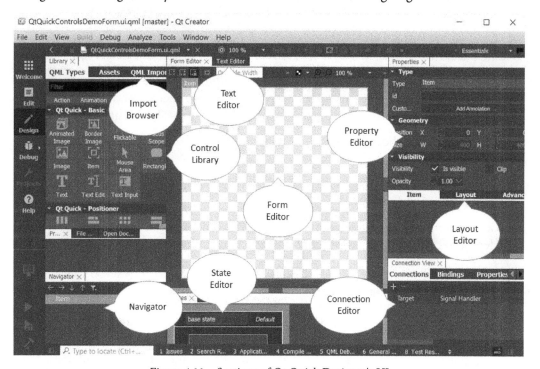

Figure 4.11 – Sections of Qt Quick Designer's UI

Qt Quick Designer's UI consists of the following major sections:

- **Navigator**: This lists the items in the current QML file as a tree structure. It's similar to the **Object Operator** window in Qt Designer that we learned about in the last chapter.

- **Control Library**: This window shows all the Qt Quick controls available in QML. You can drag and drop the controls to the canvas window to modify your UI.

- **Resources**: This displays all the resources in a list that can then be used for the UI design.

- **Import Browser**: The **Import Browser** facilitates the importing of different QML modules into the current QML file, to add new functionality to your QML project. You can also create your own custom QML module and import in from here.

- **Text Editor**: This has six tool buttons, each for a specific action, such as copy and paste.

- **Property Editor**: This is similar to the property editor of Qt Designer. The **Properties** section in Qt Quick Designer displays the properties of the selected item. You can also change the properties of the items in the **Text Editor**.

- **Form Editor**: The **Form Editor** is a canvas where you design a UI for your Qt Quick application.

- **State Editor**: This window lists the different states in a QML project, and describes UI definitions and their behavior.

- **Connection Editor**: This section is similar to the **Signal/Slot Editor** in Qt Designer. Here, you can define the signals and slots mechanism for your QML component.

You are now familiar with the Qt Quick Designer UI. Let's create a Qt Quick UI file and explore the Qt Quick controls, as follows:

1. To create a Qt Quick UI, select **QtQuick UI File** from the **New File** template screen, as shown in the following screenshot. Proceed through the next screens to create a Qt Quick form with a .ui.qml file extension. By default, Qt Creator will open up Qt Quick Designer. You can switch to code-editing mode by clicking the **Edit** button on the left-side panel:

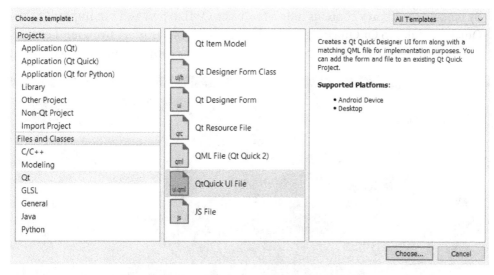

Figure 4.12 – QtQuick UI File wizard

2. Let's add a few QML elements to the **Form Editor** by grabbing a control by a mouse press and dropping it onto the canvas area by a mouse release, as shown in the following screenshot. This action is known as **drag and drop** (**DnD**). You can find several basic QML types, such as `Item`, `Rectangle`, `Image`, `Text`, and so on. `Item` is a transparent UI element that can be used as a container:

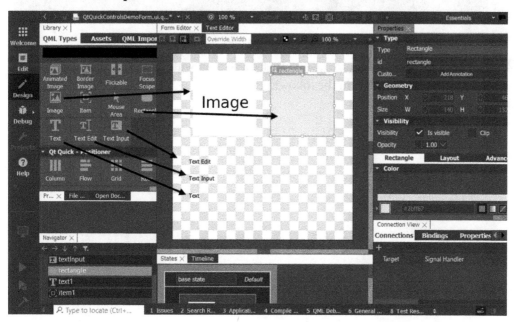

Figure 4.13 – Qt Quick Designer showing basic QML types

3. By default, the library contains only a few basic QML types. You can import Qt Quick modules to Qt Quick Designer through the QML **Import Browser**. Let's import any of the `QtQuick.Controls` packages, as shown in the next screenshot:

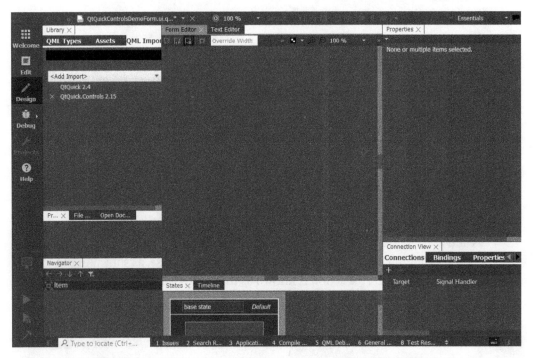

Figure 4.14 – Qt Quick Designer showing the QML module import option

4. Once the module is imported, you can see a section with **Qt Quick - Controls 2** in the library, as illustrated in the following screenshot:

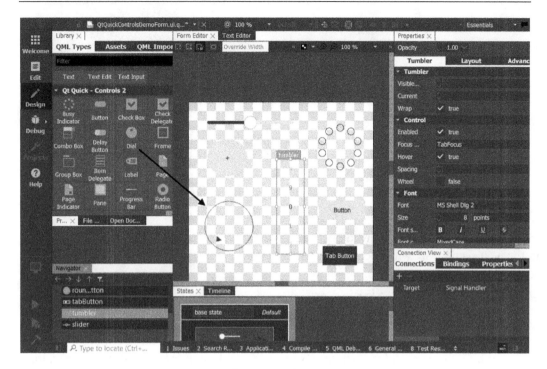

Figure 4.15 – Qt Quick Designer showing Qt Quick Controls

In this section, we got familiar with Qt Quick Designer's interfaces. In the next section, you will learn about different positioners and layouts.

Positioners and layouts in QML

There are different ways to position items in QML. You can manually position a control by mentioning *x* and *y* coordinates or by using anchors, positioners, or layouts. Let's discuss how to position a control through the aforementioned methods.

Manual positioning

A control can be positioned at specific *x* and *y* coordinates by setting their corresponding *x* and *y* properties. As per the visual coordinate system rules, this will position the controls relative to the top-left corner of their parent.

The following code snippet shows how to place a `Rectangle` item at position (50,50):

```
import QtQuick
Rectangle {
    // Manually positioned at 50,50
```

```
    x: 50 // x position
    y: 50 // y position
    width: 100; height: 80
    color: "blue"
}
```

When you run the preceding code, you will see a blue rectangle created at the (50,50) position. Change the x and y values and you will see how the position is changed relative to the top-left corner. Qt allows you to write multiple properties in a single line separated by a semicolon. You can write x and y positions in the same line, separated by a semicolon.

In this section, you learned how to position a visual item by specifying its coordinates. In the next section, we will discuss the use of anchors.

Positioning with anchors

Qt Quick provides a way to anchor a control to another control. There are seven invisible anchor lines for each item: left, right, top, bottom, baseline, horizontalCenter, and verticalCenter. You can set margins or different margins for each side. If there are multiple anchors for a specific item, they can then be grouped.

Let's have a look at the following example:

```
import QtQuick
import QtQuick.Window
Window {
    width: 400; height: 400
    visible: true
    title: qsTr("Anchoring Demo")
    Rectangle {
        id: blueRect
        anchors {
            left: parent.left; leftMargin:10
            right: parent.right; rightMargin: 40
            top: parent.top; topMargin: 50
            bottom: parent.bottom; bottomMargin: 100
        }
        color: "blue"
        Rectangle {
```

```
        id: redRect
        anchors.centerIn: blueRect
        color:"red"
        width: 150; height: 100
    }
  }
}
```

If you run this example, you will see a red rectangle inside a blue rectangle with different margins in the output window, as shown next:

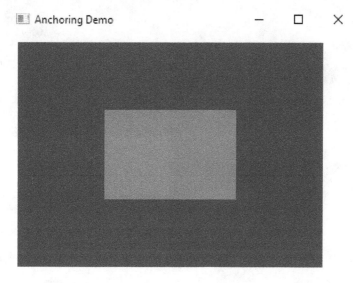

Figure 4.16 – Anchor positioning a control inside a window

In this section, you learned how to position a visual item by using anchors. In the next section, we will discuss the use of positioners.

Positioners

Positioners are containers that manage the positions of visual elements in a declarative UI. Positioners behave in a similar way to layout managers in **Qt widgets**.

A standard set of positioners is provided in a basic set of Qt Quick elements. They are outlined as follows:

- **Column** positions its children in a column.
- **Row** positions its children in a row.

- **Grid** positions its children in a grid.

- **Flow** positions its children like words on a page.

Let's have a look how to use them in Qt Quick Designer. First, create three **Rectangle** items with different colors and then position them inside a **Row** element, as illustrated in the following screenshot:

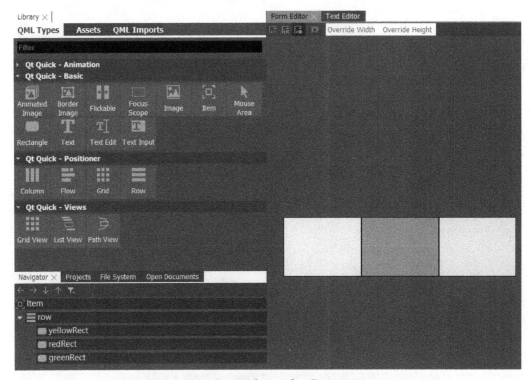

Figure 4.17 – Rectangles inside a Row positioner

You can also write code to position the controls inside a positioner. Qt Creator automatically generates code if you use Qt Quick Designer. The generated code can be viewed and modified through the **Text Editor** tab next to **Form Editor**. The code is shown in the following snippet:

```
Row {
    id: row
    Rectangle {
        id: yellowRect
        width: 150; height: 100
        color: "yellow"
```

```
        border.color: "black"
    }
    Rectangle {
        id: redRect
        width: 150; height: 100
        color: "red"
        border.color: "black"
    }
    Rectangle {
        id: greenRect
        width: 150; height: 100
        color: "green"
        border.color: "black"
    }
}
```

In this section, we learned about different positioners. In the next section, we will discuss the use of repeaters and models, along with positioners.

Repeater

A **repeater** creates a number of visual elements using a provided model, as well as elements from a template to use with a positioner, and uses data from a model. A repeater is placed inside a positioner, and creates visual elements that follow the defined positioner arrangement. When there are many similar items, then a positioner with a repeater makes it easier to maintain when arranged in a regular layout.

Let's create five rectangles positioned in a row using `Repeater`, as follows:

```
import QtQuick
import QtQuick.Window
Window {
    width: 400; height: 200
    visible: true
    title: qsTr("Repeater Demo")
    Row {
        anchors.centerIn: parent
        spacing: 10
        Repeater {
```

```
            model: 5
            Rectangle {
                width: 60; height: 40
                border{ width: 1; color: "black";}
                color: "green"
            }
        }
    }
}
```

When you run the preceding example, you will see five rectangles arranged in a row, as shown next:

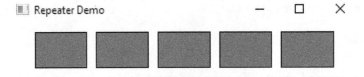

Figure 4.18 – Rectangles inside a Row positioner

In this section, we learned about the use of repeaters with positioners. In the next section, we will look into Qt Quick Layouts.

Qt Quick Layouts

Qt Quick Layouts are a set of QML types that can be used to arrange visual elements in a UI. Qt Quick Layouts can resize their children, hence they are used for resizable UIs. The basic difference between positioners and layouts is that layouts can resize their children on window resize.

Qt Quick Layouts can be imported into your QML file by using the following `import` statement:

```
import QtQuick.Layouts
```

There are five different type of layouts in QML, as outlined here:

- RowLayout: This arranges elements in a row. It is similar to GridLayout but only has one row.

- ColumnLayout: This arranges elements in a column. It is similar to GridLayout but only has one column.

- GridLayout: This allows elements to be arranged dynamically in a grid.

- Layout: This provides attached properties for items pushed onto a ColumnLayout, RowLayout, or GridLayout layout type.

- StackLayout: This arranges elements in a stack-like manner where only one element is visible at a time.

Let's look at the following RowLayout example:

```
import QtQuick
import QtQuick.Window
import QtQuick.Layouts
Window {
    width: 640; height: 480
    visible: true
    title: qsTr("Layout Demo")
    RowLayout {
        id: layout
        anchors.fill: parent
        spacing: 6
        Rectangle {
            color: 'yellow'
            Layout.fillWidth: true
            Layout.minimumWidth: 50
            Layout.preferredWidth: 150
            Layout.maximumWidth: 200
            Layout.minimumHeight: 100
            Layout.margins: 10
        }
        Rectangle {
            color: 'red'
            Layout.fillWidth: true
```

```
            Layout.minimumWidth: 50
            Layout.preferredWidth: 100
            Layout.preferredHeight: 80
            Layout.margins: 10
        }
    }
}
```

Please note that a Row type is a positioner, while a RowLayout type is a layout. When to use them depends mainly on your goal, as usual. Let's move on to the next section to see how to integrate QML with C++.

Integrating QML with C++

QML applications often need to handle more advanced and performance-intensive tasks in C++. The most common and quickest way to do this is to expose the C++ class to the QML runtime, provided the C++ implementation is derived from QObject.

QML can be easily integrated with C++ code. QML objects can be loaded and manipulated from C++. QML integration with Qt's meta-object system allows C++ functionality to be invoked from QML. This helps in building hybrid applications with a mixture of C++, QML, and JS. To expose C++ data or properties or methods to QML, it should be derived from a QObject class. This is possible because all QML object types are implemented using QObject-derived classes, allowing the QML engine to load and inspect objects through the Qt meta-object system.

You can integrate QML with C++ in the following ways:

- Embedding C++ objects into QML with context properties
- Registering the type with the QML engine
- Creating a QML extension plugin

Let's discuss each approach one by one in the following sections.

> **Important note**
>
> To quickly determine which integration method is appropriate for your project, have a look at the flowchart illustrated in the Qt documentation at the following link:
>
> https://doc.qt.io/qt-6/qtqml-cppintegration-overview.html

Embedding C++ objects into QML with context properties

You can expose C++ objects into a QML environment by using context properties. Context properties are suitable for simple applications. They export your object as a global object. Contexts are exposed to the QML environment after being instantiated by the QML engine.

Let's have a look at the following example, where we have exported radius to the QML environment. You can also export C++ models in a similar way:

```cpp
#include <QGuiApplication>
#include <QQmlApplicationEngine>
#include <QQmlContext>
int main(int argc, char *argv[])
{
    QGuiApplication app(argc, argv);
    QQmlApplicationEngine engine;
    engine.rootContext()->setContextProperty("radius", 50);
    const QUrl url(QStringLiteral("qrc:/main.qml"));
    engine.load(url);
    return app.exec();
}
```

You can use the exported value directly in the QML file, as follows:

```qml
import QtQuick
import QtQuick.Window
Window {
    width: 640; height: 480
    visible: true
    title: qsTr("QML CPP integration")
    Text {
        anchors.centerIn: parent
        text: "C++ Context Property Value: "+ radius
    }
}
```

You can also register your C++ class and instantiate it inside the QML environment. Let's learn how to achieve that in the next section.

Registering a C++ class with the QML engine

Registering QML types permits a developer to control the life cycle of a C++ object from the QML environment. This can't be achieved with context properties and also doesn't populate the global namespace. Still, all types need to be registered first and by this, all libraries need to be linked on application start, which in most cases is not really a problem.

The methods can be public slots or public methods flagged with Q_INVOKABLE. Now, let's import the C++ class into the QML file. Have a look at the following C++ class:

```cpp
#ifndef BACKENDLOGIC_H
#define BACKENDLOGIC_H
#include <QObject>
class BackendLogic : public QObject
{
    Q_OBJECT
public:
    explicit BackendLogic(QObject *parent = nullptr) {
            Q_UNUSED(parent);}
    Q_INVOKABLE int getData() {return mValue; }
private:
    int mValue = 100;
};
#endif // BACKENDLOGIC_H
```

You need to register the C++ class in the main.cpp file as a module using qmlRegisterType(), as shown here:

```cpp
qmlRegisterType<BackendLogic>("backend.logic", 1,
0,"BackendLogic");
```

Any Qobject-derived C++ class can be registered as a QML object type. Once a class is registered with the QML type system, the class can be used like any other QML type. Now, the C++ class is ready to be instantiated inside your .qml file. You have to import the module and create an object, as illustrated in the following code snippet:

```qml
import QtQuick
import QtQuick.Window
import backend.logic
Window {
    width: 640; height: 480
```

```
    visible: true
    title: qsTr("QML CPP integration")
    BackendLogic {
        id: backend
    }
    Text {
        anchors.centerIn: parent
        text: "From Backend Logic : "+ backend.getData()
    }
}
```

When you run the preceding program, you can see that the program is fetching data from the backend C++ class and displaying it in the UI.

You can also expose a C++ class as a QML singleton by using qmlRegisterSingletonType(). By using a QML singleton, you can prevent duplicate objects in the global namespace. Let's skip this part as it requires an understanding of design patterns. Detailed documentation can be found at the following link:

https://doc.qt.io/qt-6/qqmlengine.html#qmlRegisterSingletonType

In Qt 6, you can achieve C++ integration by using a QML_ELEMENT macro. This macro declares the enclosing type as available in QML, using its class or namespace name as the QML element name. To use this macro in your C++ header file, you will have to include the qml.h header file as #include <QtQml>.

Let's have a look at the following example:

```cpp
#ifndef USINGELEMENT_H
#define USINGELEMENT_H
#include <QObject>
#include <QtQml>
class UsingElements : public QObject
{
    Q_OBJECT
    QML_ELEMENT
public:
    explicit UsingElements(QObject *parent = nullptr) {
            Q_UNUSED(parent);}
    Q_INVOKABLE int readValue() {return mValue; }
```

```
private:
    int mValue = 500;
};
#endif // USINGELEMENT_H
```

In the .pro file, you have to add the qmltypes option to the CONFIG variable and QML_IMPORT_NAME and QML_IMPORT_MAJOR_VERSION are to be mentioned, as illustrated in the following code snippet:

```
CONFIG += qmltypes
QML_IMPORT_NAME = backend.element
QML_IMPORT_MAJOR_VERSION = 1
```

Your C++ class is now ready to be instantiated inside your .qml file. You have to import the module and create an object, as illustrated in the following code snippet:

```
import QtQuick
import QtQuick.Window
import backend.element
Window {
    width: 640; height: 480
    visible: true
    title: qsTr("QML CPP integration")
    UsingElements {
        id: backendElement
    }
    Text {
        anchors.centerIn: parent
        text: "From Backend Element : "+
            backendElement.readValue()
    }
}
```

In this section, you learned how to export your C++ class into the QML environment and access its functions from QML. In this example, the data is retrieved only when the method is called. You can also get notified when the data is changed inside C++ by adding a Q_PROPERTY() macro with a NOTIFY signal. You need to know about the signals and slots mechanism before using it. So, we will skip this part and discuss it further in *Chapter 6, Signals and Slots*. In the next section, we will discuss how to create a QML extension plugin.

Creating a QML extension plugin

A QML extension plugin provides the most flexible way to integrate with C++. It allows you to register types in a plugin that is loaded when the first QML file calls the import identifier. You can use plugins across projects, which is very convenient when building complex projects.

Qt Creator has a wizard to create a **Qt Quick 2 QML Extension Plugin**. Select a template, as shown in the following screenshot, and proceed with the screens that follow. The wizard will create a basic skeleton for the **QML Extension Plugin** project. The plugin class has to be derived from QqmlExtensionPlugin and should implement the registerTypes() function. A Q_PLUGIN_METADATA macro is required to identify the plugin as a QML extension plugin:

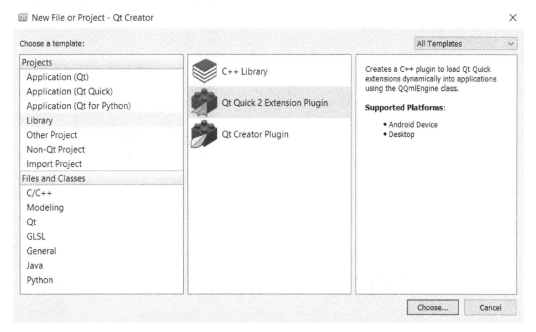

Figure 4.19 – Qt Quick 2 QML Extension Plugin wizard

This section is an advanced Qt topic. You need to understand Qt in depth to create your own QML extension plugin. You can skip this section if you are a beginner, but you can learn more about the QML extension plugin in the Qt documentation at the following link:

`https://doc.qt.io/qt-6/qtqml-modules-cppplugins.html`

Let's move on to the next section to discover how to invoke a QML method inside a C++ class.

Invoking QML methods inside a C++ class

All QML methods are exposed to the meta-object system and can be called from C++ using `QMetaObject::invokeMethod()`. You can specify types for the parameters and the return value after the colon character, as shown in the next code snippet. This can be useful when you want to connect a signal in C++ with a certain signature to a QML-defined method, for example. If you omit the types, the C++ signature will use `QVariant`.

Let's look at an application that calls a QML method using `QMetaObject::invokeMethod()`.

In the QML file, let's add a method called `qmlMethod()`, as follows:

```qml
import QtQuick
Item {
    function qmlMethod(msg: string) : string {
        console.log("Received message:", msg)
        return "Success"
    }
    Component.onCompleted: {
        console.log("Component created successfully.")
    }
}
```

In the `main.cpp` file, call `QMetaObject::invokeMethod()` as per the following code snippet:

```cpp
#include <QGuiApplication>
#include <QQmlApplicationEngine>
#include <QQmlComponent>
int main(int argc, char *argv[])
{
    QGuiApplication app(argc, argv);
```

```
    QQmlApplicationEngine engine;
    QQmlComponent component(&engine,
                            "qrc:/CustomItem.qml");
    QObject *myObject = component.create();
    QString retValue = "";
    QString msg = "Message from C++";
    QMetaObject::invokeMethod(myObject, "qmlMethod",
                              Q_RETURN_ARG(QString,
                              retValue),
                              Q_ARG(QString, msg));
    qDebug() << "QML method returned:" << retValue;
    delete myObject;
    return app.exec();
}
```

Please note that the parameter and return type have to be specified. Both basic types and object types are allowed as type names. If the type is not mentioned in the QML type system, then you must declare QVariant as a type with Q_RETURN_ARG() and Q_ARG() when calling QMetaObject::invokeMethod. Alternatively, you can call invokeMethod() with only two parameters if you don't need any return values, as shown here:

```
QMetaObject::invokeMethod(myObject, "qmlMethod");
```

In this section, you learned to receive data from a QML method. In the next section, you will learn how to access a QML object pointer inside C++.

Exposing a QML object pointer to C++

Sometimes, you may want to modify the properties of a QML object through C++, such as modifying the text of a control, changing the visibility of a control, or changing a custom property. The QML engine permits you to register your QML objects to C++ types, which automatically exposes the QML object's properties.

Let's have a look at the following example, where we have exported a QML object to the C++ environment:

```
#ifndef CUSTOMOBJECT_H
#define CUSTOMOBJECT_H
#include <QObject>
#include <QVariant>
```

```
class CustomObject : public QObject
{
    Q_OBJECT
public:
    explicit CustomObject(QObject *parent = nullptr);
    Q_INVOKABLE void setObject(QObject* object)
    {
        object->setProperty("text", QVariant("Clicked!"));
    }
};
#endif // CUSTOMOBJECT_H
```

In the QML file, you need to create an instance of the C++ class and call the C++ method. As you can see in the following code snippet, the property is manipulated inside the C++ class:

```
import QtQuick
import QtQuick.Window
import QtQuick.Controls
import MyCustomObject
Window {
    width: 640; height: 480;
    visible: true
    title: qsTr("QML Object in C++")
    CustomObject{
        id: customObject
    }
    Button {
        id: button
        anchors.centerIn: parent
        text: qsTr("Click Me!")
        onClicked: {
            customObject.setObject(button);
        }
    }
}
```

> **Important note**
>
> The Qt QML module provides several macros for registering non-instantiable types. QML_ANONYMOUS registers a C++ type that is not instantiable and cannot be referred to from QML. QML_INTERFACE registers an existing Qt interface type. The type is not instantiable from QML, and you cannot declare QML properties with it. QML_UNCREATABLE registers a named C++ type that is not instantiable but should be identifiable as a type to the QML type system. QML_SINGLETON registers a singleton type that can be imported from QML.

Congratulations! You have learned how to integrate QML and C++. In the next section, we will discuss how to use JS with QML.

Integrating QML with JS

QML has a good integration with JS and uses **JavaScript Object Notation (JSON)-like syntaxes**, allowing expressions and methods to be defined as JS functions. It also permits developers to import JS files and use the existing functionality. The QML engine provides a JS environment that has some limitations compared to the JS environment provided by a web browser. The logic for a Qt Quick application can be defined in the JS. The JS code can be written inline inside the QML file, or in a separate JS file.

Let's look at how to use inline JS inside a QML document. The following example demonstrates the btnClicked() inline JS function. The method is called when the Button control is clicked:

```
import QtQuick
import QtQuick.Window
import QtQuick.Controls
Window {
    width: 640; height: 480;
    visible: true
    title: qsTr("QML JS integration")
    function btnClicked(controlName) {
        controlName.text = "JS called!"
    }
    Column  {
        anchors.centerIn: parent
        Button {
            text:"Call JS!"
```

```
                onClicked: btnClicked(displayText)
        }
        Text {
            id: displayText
        }
    }
}
```

The preceding example shows how to integrate JS code with QML. We have used the btnClicked() inline JS function. When you run the application, you will get a message saying **JS called!**.

If your logic is very long or has uses in multiple QML documents, then use a separate JS file. You can import a JS file as follows:

```
import "<JavaScriptFile>" as <Identifier>
```

For example, you could run the following line of code:

```
import "constants.js" as Constants
```

In the previous example, we are importing constants.js into the QML environment. Constants is an identifier for our JS file.

You can also create a shared JS library. You just have to include the following line of code at the beginning of the JS file:

```
.pragma library
```

> **Important note**
>
> If the script is a single expression, then writing it inline is recommended. If the script is a few lines long, then use a block. If the script is more than several lines long or is required by different objects, then create a function and call it as needed. For long scripts, create a JS file and import it inside the QML file. Avoid using Qt.include() as it is deprecated and will be removed from future versions of Qt.

To learn more about importing JS in QML, read the following documentation:

https://doc.qt.io/qt-6/qtqml-javascript-imports.html

In this section, you learned how to integrate JS with QML. In the next section, we will discuss how to import a directory in QML.

Importing a directory in QML

You can import a local directory with QML files directly inside another QML file without adding in resources. You can use the directory's absolute or relative filesystem paths to do this, providing a convenient way for QML types to be arranged as reusable directories on the filesystem.

The common form of a directory import is shown here:

```
import "<DirectoryPath>" [as <Qualifier>]
```

For example, if your directory name is `customqmlelements`, then you can import it as follows:

```
import "../customqmlelements"
```

It is also possible to import the directory as a qualified local namespace, as shown in the following code snippet:

```
import "../customqmlelements" as CustomQMLElements
```

You can also import a file from the resource path, as follows:

```
import "qrc:/qml/customqmlelements"
```

You can also import a directory of QML files from a remote server. There are two different types of `qmldir` files: a QML directory listing file and a QML module definition file. Here, we are discussing the use of the `qmldir` QML directory listing file. The directory can be imported using the **Uniform Resource Locator** (**URL**) of the remote location. Please note that while importing over a network, only QML and JS files specified in the `qmldir` file can be accessed. To avoid malicious code, you have to be careful with the network files.

The following documentation provides further information about the `qmldir` QML directory listing file:

```
https://doc.qt.io/qt-6/qtqml-syntax-directoryimports.html
```

You can learn more about the different types of `qmldir` files at the following link:

```
https://doc.qt.io/qt-6/qtqml-modules-qmldir.html
```

In this section, you learned how to import a directory in QML. In the next section, we will discuss how to handle mouse and touch events in QML.

Handling mouse and touch events

QML provides excellent support for mouse and touch events through input handlers that let QML applications handle mouse and touch events. QML types such as MouseArea, MultiPointTouchArea, and TapHandler are used to detect mouse and touch events. We will have a look at these QML types in the following section.

MouseArea

MouseArea is an invisible item that is used with a visible item such as Item or Rectangle in order to provide mouse and touch handling events for that item. MouseArea receives mouse events within the defined area of Item. You can define this area by anchoring MouseArea to its parent's area using the anchors.fill property. If you set the visible property to false, then the mouse area becomes transparent to mouse events.

Let's look at how to use MouseArea in the following example:

```
import QtQuick
import QtQuick.Window
Window {
    width: 640; height: 480
    visible: true
    title: qsTr("Mouse Area Demo")
    Rectangle {
        anchors.centerIn: parent
        width: 100; height: 100
        color: "green"
        MouseArea {
            anchors.fill: parent
            onClicked: { parent.color = 'red' }
        }
    }
}
```

In the preceding example, you can see that only the rectangle area received the mouse event. Other parts of window didn't get the mouse events. You can perform actions accordingly based on the mouse events. MouseArea also provides convenient signals that give us information about mouse events such as mouse hover, mouse press, press and hold, mouse exit, and mouse release events. Write the corresponding signal handlers and experiment with the entered(), exited(), pressed(), and released() signals. You can also detect which mouse button was pressed and execute a corresponding action.

MultiPointTouchArea

The MultiPointTouchArea QML type enables handling of multiple touch points in a multi-touch screen. Just as with MouseArea, MultiPointTouchArea is an invisible item. You can track multiple touch points and process the gesture accordingly. When it is disabled, the touch area becomes transparent to both touch and mouse events. In a MultiPointTouchArea type, a mouse event is handled as a single touch point. You can set the mouseEnabled property to false to stop processing the mouse events.

Let's look at the following example, where we have two rectangles that follow our touch points:

```
import QtQuick
import QtQuick.Window
Window {
    width: 640; height: 480
    visible: true
    title: qsTr("Multitouch Example")
    MultiPointTouchArea {
        anchors.fill: parent
        touchPoints: [
            TouchPoint { id: tp1 },
            TouchPoint { id: tp2 }
        ]
    }
    Rectangle {
        width: 100; height: 100
        color: "blue"
        x: tp1.x; y: tp1.y
    }
    Rectangle {
```

```
        width: 100; height: 100
        color: "red"
        x: tp2.x; y: tp2.y
      }
   }
```

In a `MultiPointTouchArea` type, `TouchPoint` defines a touch point. It contains details about the touch point, such as the pressure, current position, and area. Now, run the application on your mobile device and verify it!

In this section, you learned about the use of `MouseArea` and `MultiPointTouchArea` to handle mouse and touch events. Let's learn about `TapHandler` in the next section.

TapHandler

`TapHandler` is a handler for click events on a mouse and tap events on a touchscreen. You can use `TapHandler` to react to taps and touch gestures, and it allows you to handle events in multiple nested items simultaneously. Recognition of a valid tap gesture depends on `gesturePolicy`. The default value of `gesturePolicy` is `TapHandler.DragThreshold`, for which the event point must not move significantly. If `gesturePolicy` is set to `TapHandler.WithinBounds`, then `TapHandler` takes an exclusive grab on the press event, but releases the grab as soon as the event point leaves the boundary of the parent item. Similarly, if `gesturePolicy` is set to `TapHandler.ReleaseWithinBounds`, then `TapHandler` takes an exclusive grab on the press and retains it until release in order to detect this gesture.

Let's create a `TapHandler` type that recognizes different mouse button events and stylus taps, as follows:

```
import QtQuick
import QtQuick.Window
Window {
    width: 640; height: 480
    visible: true
    title: qsTr("Hello World")
    Item {
        anchors.fill:parent
        TapHandler {
            acceptedButtons: Qt.LeftButton
            onTapped: console.log("Left Button Clicked!")
```

```
        }
        TapHandler {
            acceptedButtons: Qt.MiddleButton
            onTapped: console.log("Middle Button Clicked!")
        }
        TapHandler {
            acceptedButtons: Qt.RightButton
            onTapped: console.log("Right Button Clicked!")
        }
        TapHandler {
            acceptedDevices: PointerDevice.Stylus
            onTapped: console.log("Stylus Tap!")
        }
    }
}
```

You can use **input handlers** to handle touch events and gestures as a substitute for `MouseArea`. Input handlers make the formation of complex touch interactions simpler, which is difficult to achieve with either `MouseArea` or `TouchArea`.

Qt provides some ready-made controls to handle generic gestures such as pinch, flick, and swipe. `PinchArea` is a convenient QML type to handle simple pinch gestures. It is an invisible item that is used with another visible item. `Flickable` is another convenient QML type that provides a surface for a flick gesture. Explore the related documentation and examples to understand more about these QML elements.

Let's look at `SwipeView` in the next section.

SwipeView

A **swipe** is another common gesture in touch-based devices. You can use `SwipeView` to navigate pages by swiping sideways. It uses a swipe-based navigation model and provides a simplified way for horizontal-paged scrolling. You can add a page indicator at the bottom to display the current active page.

Let's look at a simple example, as follows:

```
import QtQuick
import QtQuick.Window
import QtQuick.Controls
```

```
Window {
    width: 640; height: 480
    visible: true
    title: qsTr("Swipe Demo")
    SwipeView {
        id: swipeView
        currentIndex: 0
        anchors.fill: parent
        Rectangle { id: page1; color: "red" }
        Rectangle { id: page2; color: "green"}
        Rectangle { id: page3; color: "blue" }
    }
    PageIndicator {
        id: pageIndicator
        count: swipeView.count
        currentIndex: swipeView.currentIndex
        anchors {
            bottom: swipeView.bottom
            horizontalCenter: parent.horizontalCenter
        }
    }
}
```

As you can see, we just have to add child items to SwipeView. You can set the SwipeView current index as the PageIndicator current index. SwipeView is one of the navigation models, which also include StackView and Drawer. You can explore these QML types to experience gestures on your mobile devices.

In this section, you learned about the use of various QML types to handle mouse, touch, and gesture events. Next, we will summarize what we learned in this chapter.

Summary

This chapter explained the fundamentals of the Qt Quick module and how to create a custom UI. You learned to design and build GUIs with Qt Quick Designer and learned about Qt Quick Controls, and how to build a custom Qt Quick application. You also learned how to integrate QML with C++ and JS. You should now understand the similarities and differences between Qt Widgets and Qt Quick and be able to choose the most suitable framework for your project. In this chapter, we have learned about Qt Quick and how to create an application using QML. You also learned how to integrate QML with JS and learned about mouse and touch events.

In the next chapter, we will discuss cross-platform development using Qt Creator. You will learn to configure and build applications on Windows, Linux, Android, and macOS **operating systems (OSes)**. We are going to learn how to port our Qt application to different platforms without too many challenges. Let's go!

Section 2: Cross-Platform Development

This section will introduce you to cross-platform development. The idea of cross-platform development is that a software application should work well on more than one platform without significant code change. This saves time in porting and maintaining the code base. This follows the Qt philosophy of "code less, create more, and deploy everywhere." In this section, you will learn about the Qt Creator IDE, its usage, and how you can develop and run the same application on different platforms.

This section includes the following chapter:

- *Chapter 5, Cross-Platform Development*

5
Cross-Platform Development

Qt has been well known for its cross-platform capability since its initial release—it was the primary vision behind creating this framework. You can use Qt Creator on your favorite desktop platforms such as Windows, Linux, and macOS, and create fluid, modern, touch-friendly **graphical user interfaces (GUIs)** and desktop, mobile, or embedded applications with the same code base or with a little modification. You can easily modify your code and deploy it on a target platform. Qt has several built-in tools to analyze your application and its performance on various supported platforms. Furthermore, it's easy to use and has an intuitive **user interface (UI)**, unlike with other cross-platform frameworks.

In this chapter, you will learn cross-platform development essentials and how to build applications on different platforms. With this, you will be able to run sample applications on your favorite desktop and mobile platforms.

In this chapter, we're going to cover the following main topics:

- Understanding cross-platform development
- Understanding compilers
- Building with qmake
- Qt project (.pro) files

- Understanding build settings
- Platform-specific settings
- Using Qt with Microsoft Visual Studio
- Running a Qt application on Linux
- Running a Qt application on macOS and iOS
- Other Qt-supported platforms
- Porting from Qt 5 into Qt 6

By the end of this chapter, you will understand Qt project files, essential settings, and how to run your Qt application on a mobile device. Let's get started!

Technical requirements

The technical requirements for this chapter include minimum versions of Qt 6.0.0 and Qt Creator 4.14.0 installed on a latest desktop platform such as Windows 10, Ubuntu 20.04, or macOS 10.14.

All the code used in this chapter can be downloaded from the following GitHub link:

```
https://github.com/PacktPublishing/Cross-Platform-Development-
with-Qt-6-and-Modern-Cpp/tree/master/Chapter05/HelloWorld
```

> **Important note**
> The screenshots used in this chapter are taken on the Windows platform. You will see similar screens based on the underlying platforms in your machine.

Understanding cross-platform development

There are several cross-platform frameworks available on the market, but Qt is a better option to select owing to its maturity and available community support. It's easy for a traditional C++ developer to adapt to Qt faster and develop high-quality applications. The Qt framework allows developers to develop applications that are compatible with multiple platforms such as Windows, Linux, macOS, **QNX** (originally known as **Quick Unix [Qunix]**), iOS, and Android. It facilitates faster application development with better code quality, with its ability to code once and its deploy-anywhere philosophy. Qt handles platform-specific implementations internally, and also enables you to build amazing ultra-lightweight applications with an impressive GUI on microcontroller-powered devices.

To develop applications using Qt for embedded platforms, you will require a commercial license to use **Qt for Device Creation**. Qt also supports some of the **microcontroller unit (MCU)** platforms such as Renesas, STM32, and NXP. At the time of writing this book, Qt for MCUs 1.8 was launched, which provides ultra-lightweight modules with a small memory footprint.

Some advantages of cross-platform development using the Qt framework are listed here:

- Cost efficiency with reduced cost of development
- Better code reusability
- Convenience
- Faster **time to market (TTM)**
- Wider market reach
- Delivers a near-native experience
- High on performance

There are also some disadvantages, such as these:

- Unavailability of platform-specific features and access to all platform **application programming interfaces (APIs)**
- Communication challenges between native and non-native components
- Certain device-specific features and hardware-compatibility challenges
- Delayed platform updates

In this section, you got a basic idea of the cross-platform nature of Qt and learned about the pros and cons of cross-platform development. Before you can run an application on any platform, you will need a compiler to compile an application for a target platform. In the next section, we will learn about compilers supported by the Qt framework.

Understanding compilers

In this section, you will learn what a compiler is and how to use it for cross-platform development. A compiler is a piece of software that transforms your program into machine code or low-level instructions that can be read and executed by a computer. These low-level machine instructions vary from platform to platform. You can compile Qt applications with different compilers such as the **GNU Compiler Collection** (GCC), or you can use a vendor-supplied one. In Qt Creator, you can find a compiler supported for a kit under the **Kits** tab, along with other essential tools for building an application on a particular platform such as Windows, Linux, or macOS. Not all supported compilers are provided with the Qt installer, but you can find the most widely used compilers automatically listed in the recommended kit. Qt may drop support for certain kit configurations or replace them with the latest version.

Currently, Qt supports the following compilers:

- GCC
- **Minimalist GNU for Windows (MinGW)**
- **Microsoft Visual C++ (MSVC)**
- **Low Level Virtual Machine (LLVM)**
- **Intel C++ Compiler (ICC)**
- **Clang** and `clang-cl`
- Nim
- QCC

Additionally, the **Qt Creator Bare Metal Device** plugin offers provision for the following compilers:

- **IAR Embedded Workbench (IAREW)**
- KEIL
- **Small Device C Compiler (SDCC)**

Apart from the preceding compilers, Qt uses specific built-in compilers while building a Qt project. These are listed here:

- **Meta-Object Compiler** (`moc`)
- **User Interface Compiler** (`uic`)
- **Resource Compiler** (`rcc`)

You can use the aforementioned compilers to build applications for a target platform or to add a custom compiler configuration. In the next section, you will learn how to create a custom compiler configuration.

Adding custom compilers

To add a compiler that is not automatically detected by Qt Creator or is unavailable, use the **Custom** option. You can specify the compiler and toolchain paths to the directories and configure these accordingly.

To add a custom compiler configuration, follow these steps:

1. To create a new compiler configuration in Qt, click on the **Tools** menu on the menu bar and then select the **Kits** tab from the left-side pane.

2. Then, click on the **Compilers** tab and select **Custom** from the **Add** dropdown. You will see **C** and **C++** options in the context menu. Select the type as per your requirement. You can see an overview of this in the following screenshot:

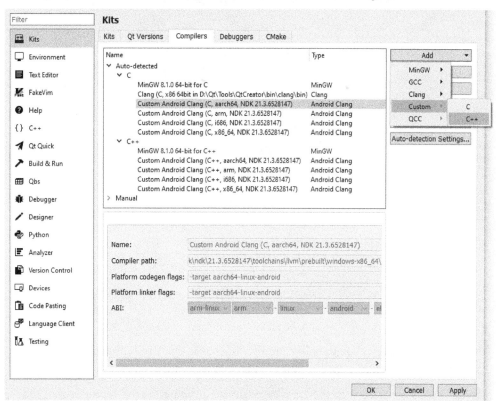

Figure 5.1 – Custom compiler option

3.   In the next step, complete the **Name** field with a customized name for the compiler.

4.   Next, in the **Compiler path** field, select a path to the directory where the compiler is located.

5.   Next, in the **Make path** field, browse a path to the directory where the make tool is located.

6.   In the next step, specify the **application binary interface (ABI)** version in the **ABI** field.

You can see an overview of this in the following screenshot:

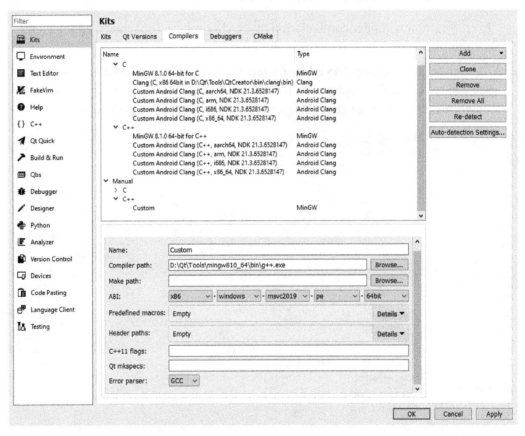

Figure 5.2 – Required fields for a custom compiler

7.   Next, you can specify the default required macros in the **Predefined macros** field. Specify each macro on separate lines in the following format: MACRO [=value].

8.   In the next step, specify in the **Header paths** field the paths to directories that the compiler checks for headers.

9.   Next, in the **C++11 flags** field, specify the flags that turn on C++11 support.

10. In the next step, specify the location of mkspecs (a set of compilation rules) in the **Qt mkspecs** field.

11. Next, in the **Error parser** field, select a suitable error parser.

12. Click on the **Apply** button to save the configuration.

In this section, you learned about supported compilers and how to create a new compiler configuration in Qt Creator, but to build and run a project we need more tools than just a compiler. Qt provides qmake as a built-in build tool for our convenience. In the next section, we will discuss what qmake is.

Building with qmake

Make is a build tool that reads project configuration file called a Makefile and builds executable programs and libraries. qmake is a Qt-provided build tool that simplifies the build process for development projects across multiple platforms. It expands the information in each project file to a Makefile that executes the necessary commands for compiling and linking. It can also be used for non-Qt projects. qmake generates a Makefile based on the information in a project file, and contains supplementary features to support development with Qt, automatically including build rules for moc and uic. qmake can also create projects for Microsoft Visual Studio without requiring the developer to change the project file.

Being a community-driven framework, Qt is really flexible toward developers and gives them the freedom to choose the most suitable tools for their project, without forcing them to use its own build system. Qt supports the following types of build systems:

- qmake
- CMake
- Qbs
- Meson
- Incredibuild

You can run qmake from the Qt Creator UI or from the command line. You should run qmake every time you make changes to your project files. Here is the syntax to run qmake from the command line:

```
>qmake [mode] [options] files
```

qmake provisions two different modes of operation. In the default mode, qmake uses the information in a project file to generate a `Makefile`, but it can also generate project files. The modes are listed as follows:

- `-makefile`
- `-project`

In **Makefile** mode, qmake will generate a `Makefile` that is used to build the project. The syntax to run qmake in Makefile mode is shown here:

```
>qmake -makefile [options] files
```

In project mode, qmake will generate a project file. The syntax to run qmake in project mode is shown here:

```
>qmake -project [options] files
```

If you use Visual Studio as an **Integrated Development Environment** (**IDE**), then you can import an existing qmake project into Visual Studio. qmake can create a Visual Studio project that contains all the essential information required by the development environment. It can recursively generate `.vcproj` files in subdirectories and a `.sln` file in the main directory, with the following command:

```
>qmake -tp vc -r
```

For example, you can generate a Visual Studio project for your `HelloWorld` project by running this command:

```
>qmake -tp vc HelloWorld.pro
```

Please note that every time you modify your project file, you need to run qmake to generate an updated Visual Studio project.

You can find more details about qmake at the following link:

https://doc.qt.io/qt-6/qmake-manual.html

Most qmake project files define the source and header files used by a project, using a list of `name = value` and `name += value` definitions, but there are additional advanced features in qmake that use other operators, functions, platform scope, and conditions to create a cross-platform application. Further details of the qmake language can be found at the following link: https://doc.qt.io/qt-6/qmake-language.html.

The Qt team has put a lot of effort into Qt 6 to make it future-proof by moving to a broadly adopted, popular build tool: **CMake**. There were changes implemented to make Qt more modular by using **Conan** as a package manager for some of the add-ons. Some of the Qt modules in Qt 6 are no longer available as binary packages in the Qt online installer but are available as Conan recipes. You can learn more about the build system changes and the addition of CMake as the default build tool at the following link: `https://doc.qt.io/qt-6/qt6-buildsystem.html`.

> **Important note**
>
> In Qt 5, the build system was made on top of qmake, but in Qt 6, CMake is the build system for building Qt from the source code. This change only affects developers who want to build Qt from sources. You can still use qmake as a build tool for your Qt applications.

In this section, you learned about qmake. We are skipping advanced qmake topics for self-exploration. In the next section, we will discuss Qt project files, which are parsed by qmake.

Qt Project (.pro) files

The `.pro` files created by Qt Creator in the earlier examples are actually Qt project files. A `.pro` file contains all the information required by qmake to build an application, a library, or a plugin. A project file supports both simple and complex build systems. A simple project file may use straightforward declarations, defining standard variables to indicate the source and header files that are used in a project. Complex projects may use multiple flow structures to optimize the build process. A project file contains a series of declarations to specify resources, such as links to the source and header files, libraries required by a project, custom-build processes for different platforms, and so on.

A Qt project file has several sections and uses certain predefined qmake variables. Let's have a look here at our earlier `HelloWorld` example `.pro` file:

```
QT          += core gui
greaterThan(QT_MAJOR_VERSION, 4): QT += widgets
CONFIG += c++17
# You can make your code fail to compile if it uses
# deprecated APIs.
# In order to do so, uncomment the following line.
#DEFINES += QT_DISABLE_DEPRECATED_BEFORE=0x060000
```

```
# disables all the APIs deprecated before Qt 6.0.0
SOURCES += \
    main.cpp \
    widget.cpp
HEADERS += \
    widget.h
FORMS += \
    widget.ui
# Default rules for deployment.
qnx: target.path = /tmp/$${TARGET}/bin
else: unix:!android: target.path = /opt/$${TARGET}/bin
!isEmpty(target.path): INSTALLS += target
```

The project file simply tells qmake what the required Qt modules in the project are, as well as the name of the executable program. It also links to the header files, source files, form files, and resource files that need to be included in the project. All of this information is crucial in order for qmake to create the configuration files and build the application. For a more complex project, you may configure your project file differently for different operating systems.

The following list describes the most frequently used variables and describes their purpose:

- QT: A list of Qt modules used in a project
- CONFIG: General project configuration options
- DESTDIR: The directory in which the executable or binary file will be placed
- FORMS: A list of UI files to be processed by the UI compiler (uic)
- HEADERS: A list of filenames of header (.h) files used when building a project
- RESOURCES: A list of resource (.qrc) files to be included in the final project
- SOURCES: A list of source code (.cpp) files to be used when building a project
- TEMPLATE: The template to use for a project

You can add different Qt modules, configurations, and definitions to your project. Let's take a look at how we can accomplish this. To add additional modules, you simply add the module keyword after QT +=, as shown here:

```
QT += core gui sql
```

You can also add a condition in front to determine when to add a specific module to your project, as follows:

```
greaterThan(QT_MAJOR_VERSION, 4): QT += widgets
```

You can also add configuration settings to your project. For example, if you want to specify c++17 specifications while compiling a project, then add the following line to your .pro file:

```
CONFIG += c++17
```

You can add a comment to a project file, starting with the hash symbol (#), and the build system will ignore the corresponding line of text. Now, let's have a look at the TEMPLATE variable. This determines whether the output of the build process will be an application, a library, or a plugin. There are different variables available to outline the type of file qmake will generate. These are listed as follows:

- app is used to build an application.

- lib is used to build a library.

- aux is used to build nothing. Use this if no compiler needs to be invoked to create a target—for instance, because your project is written in an interpreted language.

- subdirs is used for the subdirectories specified using the SUBDIRS variable. Each subdirectory must contain its own project file.

- vcapp is used to create a Visual Studio project file to build an application.

- vclib is used to create a Visual Studio project file to build a library.

- vcsubdirs is used to create a Visual Studio solution file to build projects in subdirectories.

Qt project files sometimes need to depend on the include feature. In a Qt project file, you can also define two significant variables: INCLUDEPATH and DEPENDPATH. You can use the SUBDIRS variable to compile a set of dependent libraries or modules.

Now, let's discuss what a .pri file is.

Understanding differences between .pro and .pri files

You can create a `.pri` file to include project files in a complex project. This improves readability and segregates different modules. A `.pri` file is usually called a **project include file** or a qmake include file, and its format is similar to that of a `.pro` file. The main difference is in the intent of use; a `.pro` file is what we expect to run qmake on directly, while a `.pri` file is included by a `.pro` file. You can add common configurations such as source files, header files, `.ui` files, and `.qrc` files into `.pri` files and include them from multiple `.pro` files as per your project needs.

You can include a `.pri` file inside a `.pro` file, as illustrated here:

```
include($$PWD/common.pri)
```

In this section, you learned about what a Qt project file is, as well as the different variables used in it. In the next section, we will discuss different build settings.

Understanding build settings

Before a project is compiled or built, the compiler requires certain details, which are known as the build settings. This is a very important part of the compilation process. In this section, you will learn about build settings and how to configure them in a proper way. You can have multiple build configurations for the same project. Usually, Qt Creator creates debug, release, and profile build configurations automatically. A debug build contains additional debug symbols required for debugging an application, whereas the release version is an optimized version without such symbols. Generally, developers use a debug configuration for testing and a release configuration for creating the final binaries. A profile build is an optimized release build that is delivered with separate debug information and is best suited to analyzing applications.

Build settings can be specified in the **Projects** mode. You may find that the **Projects** button is disabled if there are no projects opened in the IDE. You can add a new build configuration by clicking the **Add** drop-down button and then selecting the type of configuration you would like to add. The options may depend on the build system selected for the project. You can add multiple build configurations as per your requirement. You can click on the **Clone...** button to add a build configuration based on the current build configuration, or click on the **Rename...** button to rename the currently selected build configuration. Click on the **Remove** button to remove a build configuration.

You can see an overview of this in the following screenshot:

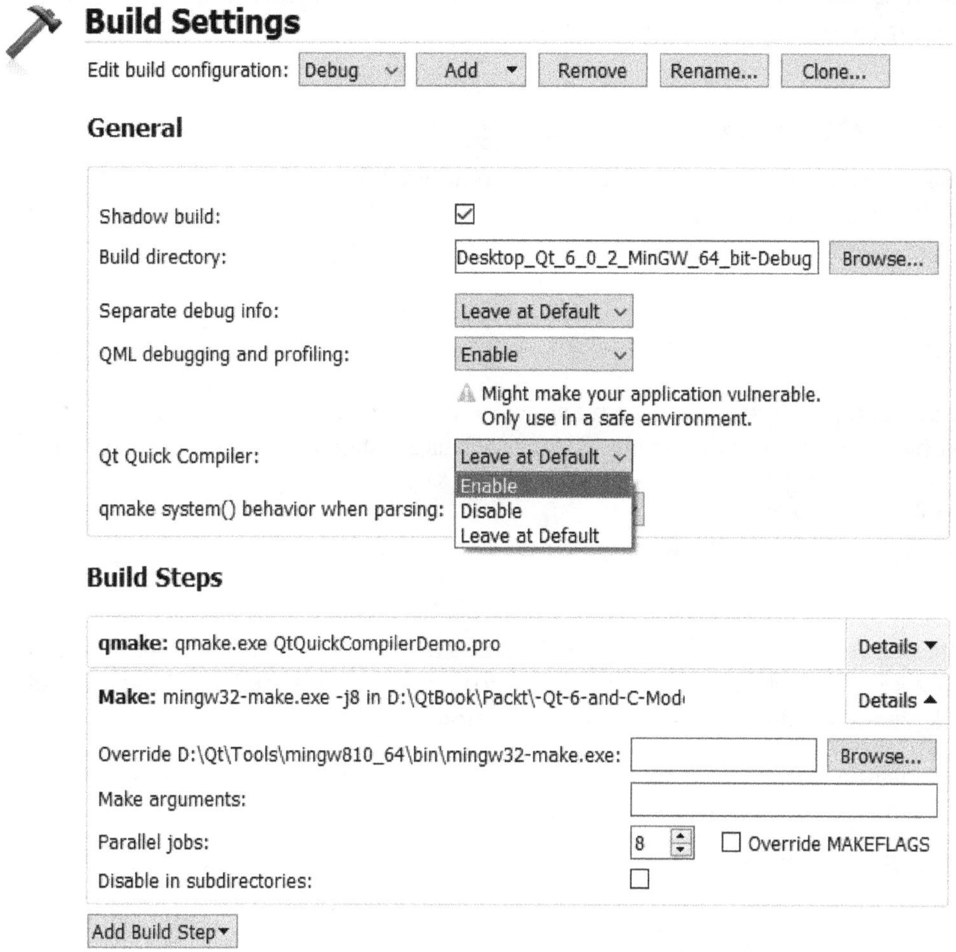

Build Settings

Edit build configuration: Debug ˅   Add ▾   Remove   Rename...   Clone...

General

Shadow build: ☑

Build directory: Desktop_Qt_6_0_2_MinGW_64_bit-Debug   Browse...

Separate debug info: Leave at Default ˅

QML debugging and profiling: Enable   ˅

⚠ Might make your application vulnerable.
Only use in a safe environment.

Qt Quick Compiler: Leave at Default ˅
                                                 Enable
qmake system() behavior when parsing: Disable
                                                 Leave at Default

Build Steps

qmake: qmake.exe QtQuickCompilerDemo.pro   Details ▾

Make: mingw32-make.exe -j8 in D:\QtBook\Packt\-Qt-6-and-C-Mod   Details ▲

Override D:\Qt\Tools\mingw810_64\bin\mingw32-make.exe: [   ]   Browse...

Make arguments: [   ]

Parallel jobs: 8 ⬍   ☐ Override MAKEFLAGS

Disable in subdirectories: ☐

Add Build Step▾

Figure 5.3 – Build settings and Qt Quick Compiler option

Normally, Qt Creator builds projects in a different directory from the source directory, known as shadow builds. This segregates the files generated for each build and run kit. If you want to only build and run with a single kit, then you can deselect the **Shadow build** checkbox. The Qt Creator project wizard creates a Qt Quick project that can be compiled to use the **Qt Resource System**. To use the default settings, select **Leave at Default**. To compile Qt Quick code, select **Enable** in the **Qt Quick Compiler** field, as shown in *Figure 5.3*.

You can read more about different build configurations at the following link:

`https://doc.qt.io/qtcreator/creator-build-settings.html`

In this section, we discussed build settings. While building a cross-platform application, it is important to add platform-specific configurations to the project file. In the next section, we will learn about platform-specific settings.

Platform-specific settings

You can define different configurations for different platforms, since not every configuration can fit all use cases. For example, if you want to include different header paths for different operating systems, you can add the following lines of code to your `.pro` file:

```
win32: INCLUDEPATH += "C:/mylibs/windows_headers"
unix:INCLUDEPATH += "/home/user/linux_headers"
```

In the preceding code snippet, we have added some Windows-specific and Linux-specific header files. You can also put configurations such as `if` statements in C++, as shown here:

```
win32 {
    SOURCES += windows_code.cpp
}
```

The preceding code is intended only for Windows platforms, which is why we have added a `win32` keyword before it. If your target platform is based on Linux, then you can add a `unix` keyword to add Linux-specific configurations.

To set a custom icon for your application on the Windows platform, you should add the following line of code to your project (`.pro`) file:

```
RC_ICONS = myapplication.ico
```

To set a custom icon for your application on macOS, you should add the following line of code to your project (`.pro`) file:

```
ICON = myapplication.icns
```

Note that the icon format is different for Windows and macOS. For Linux distributions, there is a different approach to making the desktop entry for each flavor.

In this section, we discussed some of the platform-specific settings. In the next section, we will learn about the use of Visual Studio with Qt VS Tools.

Using Qt with Microsoft Visual Studio

Some developers choose Visual Studio as their preferred IDE. So, if your favorite IDE is Visual Studio, then you can integrate Qt VS Tools with Microsoft Visual Studio. This will allow you to use the standard Windows development environment without having to worry about Qt-related build steps or tools. You can install and update Qt VS Tools directly from Microsoft Visual Studio.

You can find Qt Visual Studio Tools from Visual Studio Marketplace for the corresponding versions. For Visual Studio 2019, you can download the tool from the following link: `https://marketplace.visualstudio.com/items?itemName=TheQtCompany.QtVisualStudioTools2019`. You can also download the VS add-in from the following Qt download link: `https://download.qt.io/official_releases/vsaddin/`.

These are some of the important features of Qt VS Tools:

- Wizards to create new projects and classes
- Automated build setup for `moc`, `uic`, and `rcc` compilers
- Import and export of `.pro` and `.pri` files
- Automatic conversion of a Qt VS Tools project to a `qmake` project
- Integrated Qt resource management
- Ability to create Qt translation files and integration with **Qt Linguist**
- Integrated **Qt Designer**
- Integrated Qt documentation
- Debugging extensions for Qt data types

To start using the features in the Visual Studio environment, you must set the Qt version. Select the appropriate version from **Options** and restart the IDE. Visual Studio and Qt use different file formats to save projects. You may use `.pro` files with `qmake` or `.vcproj` files with Visual Studio to build your project. Since Visual Studio is used for Windows-specific development, it is recommended to use Qt Creator as the IDE for cross-platform development.

If you don't have a `.vcproj` file, then you can generate one from a `.pro` file through the command line or through VS Tools. We have already discussed the command-line instruction in the *Building with qmake* section. You can also convert your `.pro` file to a `.vcproj` file by using the **Open** option in VS Tools. Please note that the generated `.vcproj` file only contains Windows-specific settings.

In this section, we discussed the VS add-in. In the next section, we will learn how to run a sample application on Linux. We will skip a discussion on building and running a Qt application on Windows as we have already discussed this in earlier chapters.

Running a Qt application on Linux

Building and running a Qt application on Linux is similar to running it on Windows, but Linux has many distributions and thus it is difficult to build an application that flawlessly runs on all Linux variants. In most distributions, the application will run smoothly. We will focus on Ubuntu 20.04 as our target platform. When you install Qt on Ubuntu, it will automatically detect the kit and configurations automatically. You can also configure a kit with the appropriate compiler and Qt version, as illustrated in the following screenshot:

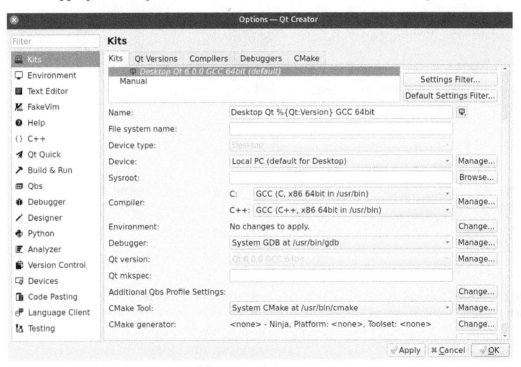

Figure 5.4 – Desktop kit configuration on Ubuntu

Let's run our HelloWorld example on Ubuntu. Hit the **Run** button on the left-side pane. A UI showing **Hello World!** will appear in no time, as illustrated in the following screenshot:

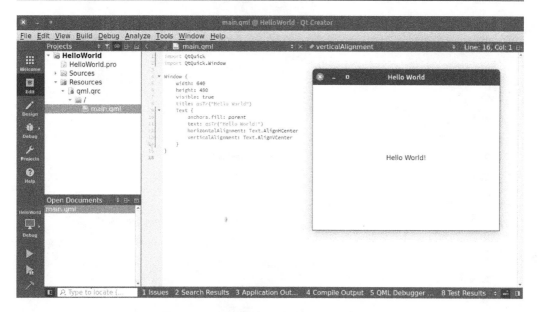

Figure 5.5 – Application running on Ubuntu

You can also run the application from the command line, as shown in the following code snippet:

```
$./HelloWorld
```

In this section, we discussed how to run our application on Linux distributions. In the next section, we will learn about running a Qt application on macOS and iOS.

Running a Qt application on macOS and iOS

We have already discussed how to build and run applications on Windows and Linux platforms in earlier chapters. Let's move on to learn how to run our applications on platforms such as macOS and iOS. To build a Qt application on macOS and iOS, you will need Xcode from the App Store. Xcode is the IDE for macOS, comprising a suite of software development tools for developing applications in macOS and iOS. If you have already installed Xcode, Qt Creator will detect its existence and will automatically detect the suitable kits. As for the kit selection, Qt for macOS supports kits for Android, `clang` 64-bit, iOS, and iOS Simulator.

You can see a sample desktop kit configuration on macOS in the following screenshot:

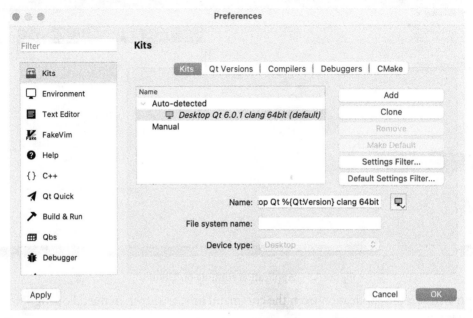

Figure 5.6 – Desktop kit configuration on macOS

You can also manually add a debugger in the **Debuggers** tab if you don't want to use the **Auto-detected** debugger, as illustrated in the following screenshot:

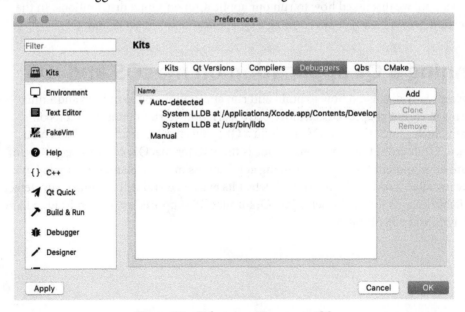

Figure 5.7 – Debugger option on macOS

Running an application on macOS is similar to running it on Windows. Just hit the **Run** button and you will see the application running in no time.

Mobile platforms hold equal importance to desktop platforms such as Windows, Linux, and macOS. Let's explore how to set up an environment for running applications on iOS.

Configuring Qt Creator for iOS

Running Qt applications on iOS is really simple. You can connect your iOS device and select a suitable device type from the device selection list. You can select **Device type** from the **Kits** selection screen. You can also run the application on iOS Simulator, as illustrated in the following screenshot:

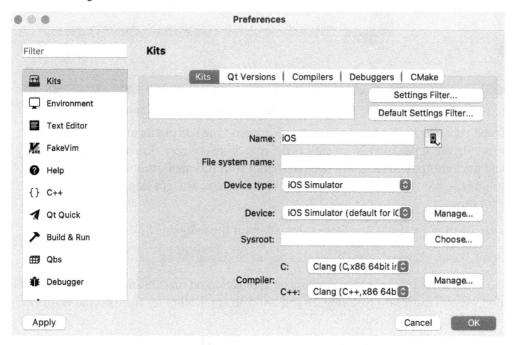

Figure 5.8 – iOS Simulator option on macOS

After configuring the kit, just plug in your iPhone and hit the **Run** button. You can see a sample output in the following screenshot:

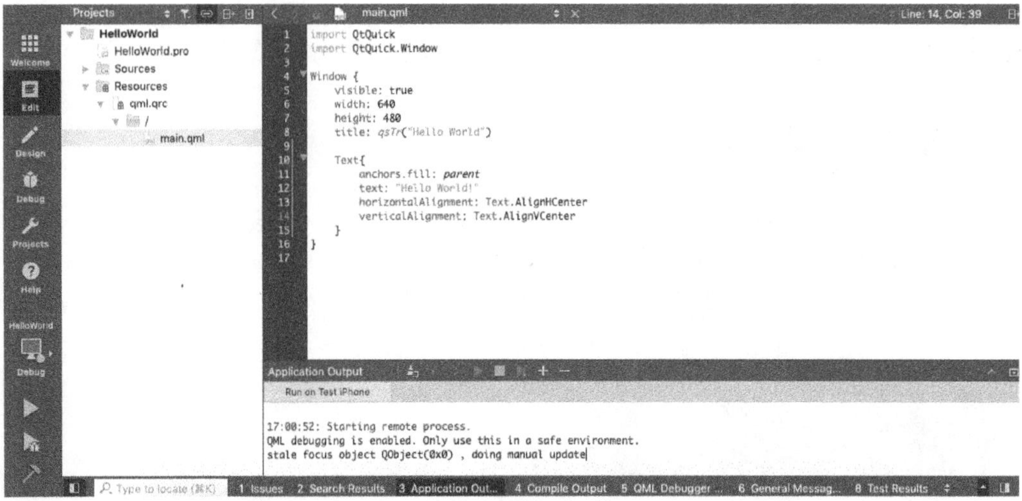

Figure 5.9 – Qt Creator running an application on an iPhone

It is relatively easy to build and run an application on the iOS platform. However, distributing the application is not easy as the App Store is a very closed ecosystem. You should have an Apple ID and will need to sign in your iOS applications before you can distribute them to your users. You can't avoid these steps, but let's skip the deployment part for now.

You can learn more about App Store submissions at the following link:

`https://developer.apple.com/app-store/submissions`

In this section, we learned about running an application on macOS and iOS. In the next section, we will learn how to configure and build an application for the Android platform.

Configuring Qt Creator for Android

Android is the most popular mobile platform today, hence developers want to build applications for Android. Although Android is a Linux-based operating system, it is very different from other Linux distributions. In order to use it, you have to configure Qt Creator and install certain packages.

For smooth functioning of your Qt Creator configuration for Android, use OpenJDK 8, NDK r21 with clang toolchain. You can run sdkmanager from the `ANDROID_SDK_ROOT\cmdline-tools\latest\bin` with required arguments to configure with required dependencies.

You can learn more about android specific requirements and instructions in the following link:

`https://doc.qt.io/qt-6/android-getting-started.html`

Let's get started with configuring your machine for Android by following these next steps:

1. To build a Qt application on Android, you have to install the Android **software development kit (SDK)**, the Android **native development kit (NDK)**, the **Java Development Kit (JDK)**, and OpenSSL to your development PC, irrespective of your desktop platform. You will find the download option with a globe icon or **Download** button next to each corresponding field, to download from the respective package's page.

2. After all the required packages are installed, restart Qt Creator. Qt Creator should be able to detect the build and platform tools automatically.

3. However, you may have to configure further to fix errors in **Android** settings. You may find the SDK manager, the platform SDK, and essential packages missing, as shown in the following screenshot:

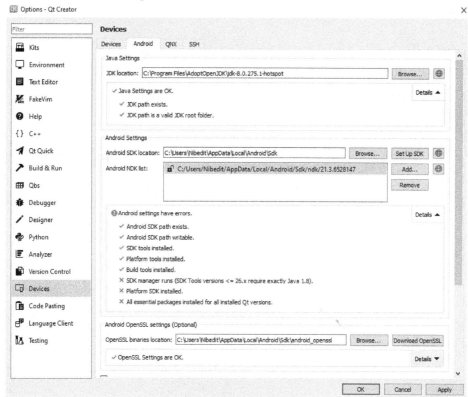

Figure 5.10 – Android Settings screen

4. Select the correct SDK and NDK path under **Android Settings**. Click on the **Apply** button to save the changes.

5. Click on the **SDK Manager** tab and click on the **Update Installed** button. You may see a message box prompting you to install missing packages, as illustrated in the following screenshot. Click on the **Yes** button to install the packages:

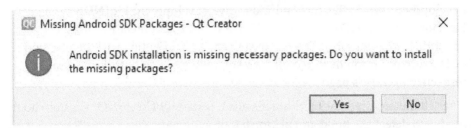

Figure 5.11 – Information message showing missing Android packages

6. You may get another message warning of Android SDK changes, listing missing essential packages, as illustrated in the following screenshot. Click on the **OK** button:

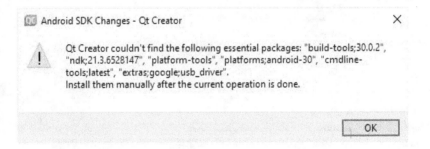

Figure 5.12 – Warning about missing Android packages

7. Click on the **Advanced Options...** button to launch the **SDK Manager Arguments** screen, type `--verbose`, and click on the **OK** button. You can see an overview of this in the following screenshot:

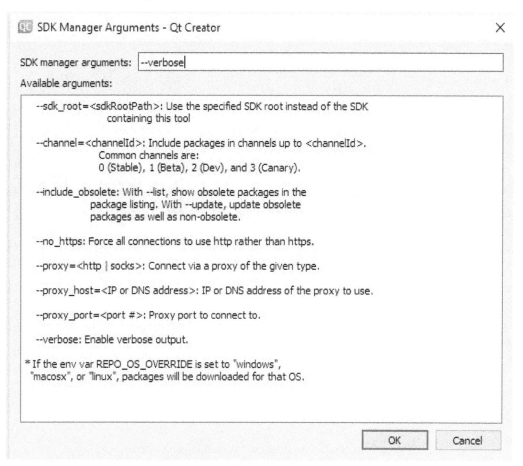

Figure 5.13 – Android SDK Manager tool

8. Once the issues are resolved, you will see that all Android settings have been properly configured, as shown in the following screenshot:

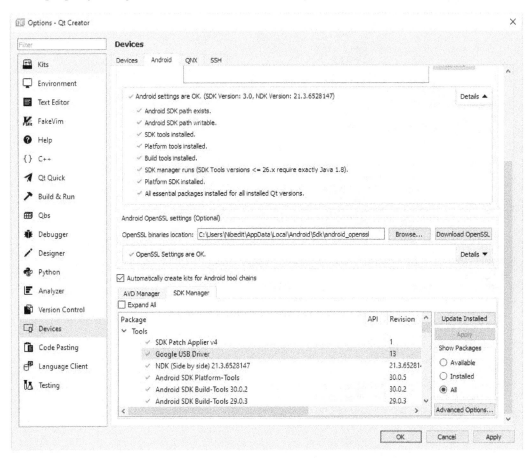

Figure 5.14 – Proper Android configuration in Qt Creator

9. If the issues are still not resolved or if you want to install a specific platform, you can enter the appropriate command, as shown in the following screenshot. You may also install the required packages from the command line. Qt will automatically detect the build tools and platforms available in the SDK location:

Figure 5.15 – Android SDK Manager tool

10. Once the Android settings are properly configured, you can see the Android kit is ready for development, as illustrated in the following screenshot:

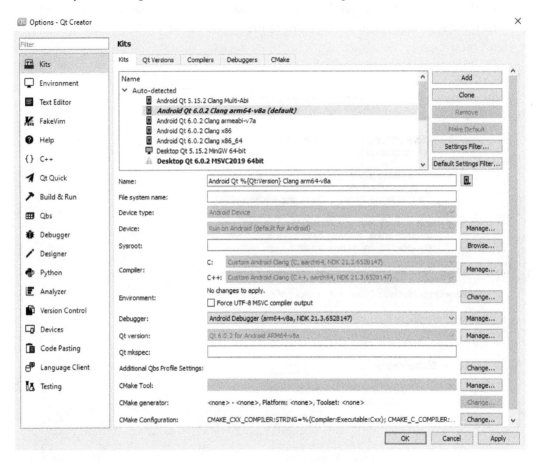

Figure 5.16 – Properly configured Android kit

11. Select an Android kit from the **Kit** selection option, as illustrated in the following screenshot:

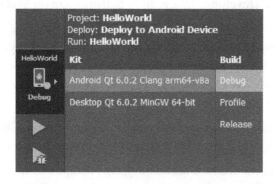

Figure 5.17 – Android Kit selection option

12. In this step, you can select a target Android version and configure your Android application by creating a `AndroidManifest.xml` file with Qt Creator. You can set the package name, version code, SDK version, application icon, permissions, and so on. The settings can be seen in the following screenshot:

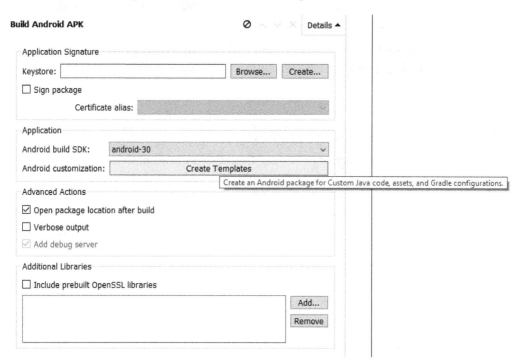

Figure 5.18 – Android manifest option in build settings

13. Your machine is now ready for Android development. However, your Android hardware requires developer options to be enabled or the Android emulator to be used. To enable the **Developer** mode, go to **Settings**, tap on **System**, and then on **About phone**.

14. Then, tap on **Software info** and find the build number. Keep tapping **Builder number** until you see **Developer** mode activated. It may take seven taps to activate the **Developer** mode. Now, go back to the **Settings** pane, where you will now find **Developer** options as an entry.

15. Your Android device is ready to run the Android application. Click on the **Run** button and select a device from the **Compatible device** list screen.

16. Next, tap **Allow** on the **Allow USB Debugging** prompt on the Android device. You will see the **Hello World!** message running on your Android device. You can find the .apk file generated inside the build folder.

Congratulations! You have successfully developed your Android application. Unlike iOS, Android is an open system. You can copy or distribute the .apk file into other Android devices running on the same Android version, and then install it. However, if you want to distribute your apps on Google Play Store, then you will have to register as a Google Play developer and sign the package.

In this section, we learned how to configure and build for an Android platform. In the next section, we will discuss other platforms supported by Qt 6 at the time this book was authored.

Other Qt-supported platforms

Qt 5 had support for a great range of platforms, from desktop and mobile platforms to embedded and web platforms. Qt 6 is yet to support all platforms that were supported in Qt 5, but the platforms will be gradually supported as Qt 6 matures. Currently, only embedded Linux is supported in the latest release of Qt 6 under the commercial license. You may have to wait some time to port your application to Qt 6 on a different embedded platform. Otherwise, if you want to migrate to Qt 6 immediately for your favorite embedded platform, you have to build from the source code and do the necessary modifications.

The following link provides a snapshot of embedded Linux support in Qt 6.2: https://doc-snapshots.qt.io/qt6-dev/embedded-linux.html. This link may get updated as Qt moves to the next release.

Qt also provides a **Boot to Qt** software stack for embedded Linux systems under commercial licenses. It is a lightweight, Qt-optimized complete software stack that is installed on the target system. The conventional embedded Linux kernel is used in the Boot to Qt software stack, which is designed with the Poky and Yocto packages.

Explore more about Boot to Qt at the following link:

```
https://doc.qt.io/QtForDeviceCreation/b2qt-index.html
```

Qt for WebAssembly allows you to build Qt applications for web platforms. It does not necessarily require any client-side installations, and saves server resources. It is a platform plugin that lets you build Qt applications that can be embedded into web pages. It is not yet available to open source developers in Qt 6. Commercial license holders may get early access to use this plugin.

You can learn more about the **Qt for WebAssembly** plugin at the following link:

```
https://wiki.qt.io/Qt_for_WebAssembly
```

In this section, we learned about other platforms supported in Qt 6. In the next section, we will discuss how to port your application from Qt 5 to Qt 6.

Porting from Qt 5 into Qt 6

Qt 6 is a major change to the Qt framework, therefore it breaks some of the backward compatibility. So, before upgrading to Qt 6, make sure that your Qt 5 application is updated to Qt 5.15. Porting will be easier from Qt 5.15 to Qt 6, with the fewest number of changes. However, APIs marked as deprecated or obsolete in Qt 5.15 may have been removed from Qt 6.0.

The CMake APIs in Qt 5 and Qt 6 are almost identical in terms of semantics. As a result, Qt 5.15 introduced versionless targets and commands, allowing CMake code to be written that is completely independent of Qt versions. Versionless imported targets are most useful for projects that require both Qt 5 and Qt 6 compilation. It is not recommended to use them by default because of the missing target properties. You can read more on this at the following link: `https://doc.qt.io/qt-6/cmake-qt5-and-qt6-compatibility.html`.

Some of the classes and modules have been removed in Qt 6, but these are kept in `Qt5Compat` for ease of porting. Apart from build system changes, you may need to fix up the includes directives of obsolete classes—for example, classes such as `QLinkedList`, `QRegExp`, and `QTextCodec` are replaced in Qt6 with new classes. But for ease of porting, you need to add `core5compat` into your `.pro` file, as shown here:

```
QT += core5compat
```

There are also changes with respect to the drawing mechanism. If you were using OpenGL-style **OpenGL Shading Language** (**GLSL**) in your project, then you would have to switch to Vulkan-style GLSL. As per new changes, you can write shaders in Vulkan-compatible GLSL and use the qsb tool. Your shader code should be compiled into **Standard Portable Intermediate Representation-Vulkan** (**SPIR-V**) format. We will discuss graphics in detail in *Chapter 8, Graphics and Animations*. Further details can be found at the following link: `https://doc.qt.io/qt-6/qtshadertools-index.html`.

There are also some changes to **Qt Modeling Language** (**QML**). The Qt Quick Extras module has merged with Qt Quick Controls. Modules such as `QtGraphicalEffects` have been removed from Qt 6 and will be available with a different license. Qt Quick MultiEffect is available in the Qt Marketplace and provides better performance. You might also consider updating your earlier signal connections in QML to use a JavaScript function declaration, as shown in the following code snippet:

```
Connections {
    target: targetElement
    function onSignalName() {//Do Something}
}
```

The Qt State Machine module is largely source-compatible with the Qt 5 version, so you should be able to continue working on their projects with no—or only slight—changes. To use the State Machine module's classes, add the following line of code to your Qt project (`.pro`) file:

```
QT += statemachine
```

To import the State Machine module inside the QML file, use the following `import` statement:

```
import QtQml.StateMachine
```

Qt provides detailed porting guidelines. Have a look at the following documentation if you are looking to port your Qt 5 applications to Qt 6:

`https://doc.qt.io/qt-6/portingguide.html`

`https://www.qt.io/blog/porting-from-qt-5-to-qt-6-using-qt5compat-library`

`https://doc.qt.io/qt-6/porting-to-qt6-using-clazy.html`

In this section, you learned how to port your application from Qt 5 to Qt 6. In the next section, we will summarize what we learned in this chapter.

Summary

This chapter explained cross-platform development using Qt Creator. You learned about various compilers, build tools, and build- and platform-specific settings. In this chapter, you learned to configure and build applications on desktop and mobile platforms and how to run applications on iPhone and Android devices. We discussed how to port your Qt project to different platforms without too many challenges.

In the next chapter, you will learn about the signal and slots mechanism, the Qt meta object system, and event handling. Let's continue!

Section 3: Advanced Programming, Debugging, and Deployment

In this section, you will learn about advanced programming and development methodologies. You will learn about debugging, testing, and deploying Qt applications on various platforms. You will also learn about internationalization and how to build high-performance applications.

In this section, there are the following chapters:

- *Chapter 6, Signals and Slots*
- *Chapter 7, Model View Programming*
- *Chapter 8, Graphics and Animations*
- *Chapter 9, Testing and Debugging*
- *Chapter 10, Deploying Qt Applications*
- *Chapter 11, Internationalization*
- *Chapter 12, Performance Considerations*

6
Signals and Slots

In the previous chapters, we learned how to create GUI applications with Qt Widgets and Qt Quick. But to make our applications usable, we need to add a communication mechanism. The **signals** and **slots** mechanism is one of the distinct features of Qt and makes it unique from other frameworks. Signals and slots are implemented through Qt's meta-object system.

In this chapter, you will learn about signals and slots in depth and how they work internally. You will be able to receive notifications from different classes and take the corresponding action.

In this chapter, we will discuss the following topics:

- Understanding Qt signals and slots
- The working mechanism of Qt signals and slots
- Getting to know Qt's property system
- Understanding signals and the handler event system
- Understanding events and the event loop
- Managing events with an event filter
- Drag and drop

By the end of this chapter, you will be able to communicate between C++ classes with QML and between QML components.

Technical requirements

The technical requirements for this chapter include having the minimum versions of Qt (6.0.0) and Qt Creator (4.14.0) installed on the latest desktop platform available, such as Windows 10, Ubuntu 20.04, or macOS 10.14.

All the code in this chapter can be downloaded from the following GitHub link:

```
https://github.com/PacktPublishing/Cross-Platform-Development-
with-Qt-6-and-Modern-Cpp/tree/master/Chapter06
```

> **Important note**
>
> The screenshots in this chapter were taken on a Windows machine. You will see similar screens based on the underlying platforms on your machine.

Understanding Qt signals and slots

In GUI programming, when a user performs any action with any UI element, another element should get updated, or a certain task should be done. To achieve this, we want communication between objects. For example, if a user clicks the **Close** button on the **Title** bar, it is expected that the window closes. Different frameworks use different approaches to achieve this kind of communication. A **callback** is one of the most commonly used approaches. A callback is a function that's passed as an argument to another function. Callbacks can have multiple drawbacks and may suffer from complications in ensuring the type-correctness of callback arguments.

In the Qt framework, we have a substitute for this callback technique known as signals and slots. A signal is a message that is passed to communicate that the state of an object has changed. This signal may carry information about the change that has occurred. A slot is a special function that is invoked in response to a specific signal. Since slots are functions, they contain logic to perform a certain action. Qt Widgets have many predefined signals, but you can always extend your classes and add your own signals to them. Similarly, you can also add your own slots to handle the intended signal. Signals and slots make it easy to implement the observer pattern while avoiding boilerplate code.

To be able to communicate, you must connect the corresponding signals and slots. Let's understand the connection mechanism and syntaxes of a signal and slot connection.

Understanding syntax

To connect a signal to a slot, we can use `QObject::connect()`. This is a thread-safe function. The standard syntax is as follows:

```
QMetaObject::Connection QObject::connect(
    const QObject *senderObject, const char *signalName,
    const QObject *receiverObject, const char *slotName,
    Qt::ConnectionType type = Qt::AutoConnection)
```

In the preceding connection, the first argument is the sender object, while the next argument is the signal from the sender. The third argument is the receiver object, while the fourth is the slot method. The last argument is optional and describes the type of connection to be established. It determines whether the notification will be delivered to the slot immediately or queued for later. There are six different types of connections that can be made in Qt 6. Let's have a look at the connection types:

- **Qt::AutoConnection**: This is the default type of connection. This connection type is determined when the signal is emitted. If both the sender and receiver are in the same thread, then `Qt::DirectConnection` is used; otherwise, `Qt::QueuedConnection` is used.

- **Qt::DirectConnection**: In this case, both the signal and slot live in the same thread. The slot is called immediately after the signal is emitted.

- **Qt::QueuedConnection**: In this case, the slot lives in another thread. The slot is called once control returns to the event loop of the receiver's thread.

- **Qt::BlockingQueuedConnection**: This is similar to `Qt::QueuedConnection`, except that the signaling thread blocks until the slot returns. This connection must not be used if both the sender and receiver are in the same thread to avoid deadlock.

- **Qt::UniqueConnection**: This can be combined with any one of the aforementioned connection types, using a `bitwise OR`. This is used to avoid duplicate connections. The connection will fail if the connection already exists.

- **Qt::SingleShotConnection**: In this case, the slot is called only once and the connection is disconnected once the signal is emitted. This can be also used with other connection types. This connection type was introduced in Qt 6.0.

> **Important note**
>
> You must be careful while using `Qt::BlockingQueuedConnection` to avoid deadlocks. You are sending an event to the same thread and then locking the thread, waiting for the event to be processed. Since the thread is blocked, the event will never be processed, and the thread will be blocked forever, causing a deadlock. Use this connection type if you know what you are doing. You must know the implementation details of both threads before using this connection type.

There are several ways to connect signals and slots. You must use the `SIGNAL()` and `SLOT()` macros when specifying the signal and the slot function, respectively. The most commonly used syntax is as follows:

```
QObject::connect(this, SIGNAL(signalName()),
                 this, SLOT(slotName()));
```

This is the original syntax that has been around since the beginning of Qt. However, its implementation has changed quite a few times. New features have been added without breaking the basic **Application Programming Interface (API)**. It is recommended to use the new function pointer syntax, as shown here:

```
connect(sender, &MyClass::signalName, this,
        &MyClass::slotName);
```

There are pros and cons to both syntaxes. You can learn more about the differences between **string-based** and **functor-based** connections at

`https://doc.qt.io/qt-6/signalsandslots-syntaxes.html`

If a connection fails, then the preceding statement returns `false`. You can also connect to functors or C++11 lambdas, as follows:

```
connect(sender, &MyClass::signalName, this, [=]()
        { sender->doSomething(); });
```

You can check the return value to verify whether the signal connected to the slot successfully. The connection can fail if the signatures aren't compatible, or the signal and slot are missing.

> **Important note**
>
> `Qt::UniqueConnection` does not work for lambdas, non-member functions, and functors; it can only be used to connect to member functions.

The signatures of signals and slots may contain arguments, and these arguments may have default values. You can connect a signal to a slot if the signal has at least as many arguments as the slot, as well as if there is a possible implicit conversion between the types of the corresponding arguments. Let's look at feasible connections with varying numbers of arguments:

```
connect(sender, SIGNAL(signalName(int)), this,
        SLOT(slotName(int)));
connect(sender, SIGNAL(signalName(int)), this,
        SLOT(slotName()));
connect(sender, SIGNAL(signalName()), this,
        SLOT(slotName()));
```

However, the following one won't work as the slot has more arguments than the signal:

```
connect(sender, SIGNAL(signalName()), this,
        SLOT(slotName(int)));
```

Every connection you make emits a signal, so duplicate connections emit two signals. You can break a connection using `disconnect()`.

You can also use Qt with a third-party signal/slot mechanism. If you want to use both mechanisms for the same project, then add the following configuration to your Qt project (`.pro`) file:

```
CONFIG += no_keywords
```

Let's create an example with a simple signal and slot connection.

Declaring signals and slots

To create a signal and slot, you must declare the signal and slot inside your custom class. The header file of the class will look like this:

```
#ifndef MYCLASS_H
#define MYCLASS_H
#include <QObject>
class MyClass : public QObject
{
    Q_OBJECT
public:
```

```
    explicit MyClass(QObject *parent = nullptr);
signals:
    void signalName();
public slots:
    void slotName();
};
#endif // MYCLASS_H
```

As you can see, we have added Q_OBJECT to the class to facilitate the signals and slots mechanism. You can declare a signal with the signals keyword in your header file, as shown in the previous snippet. Similarly, slots can be declared with the slots keyword. Both signals and slots can have arguments. In this example, we used the same object for the sender and receiver to make this explanation simpler. In most cases, signals and slots will be located in different classes.

Next, we will discuss how to connect the signal to the slot.

Connecting the signal to the slot

Previously, we declared a custom signal and slot. Now, let's look at how to connect them. You can define a signal and slot connection and emit the signal inside MyClass, as follows:

```
#include "myclass.h"
#include <QDebug>
MyClass::MyClass(QObject *parent) : QObject(parent)
{
    QObject::connect(this, SIGNAL(signalName()),
            this, SLOT(slotName()));
    emit signalName();
}
void MyClass::slotName()
{
    qDebug() << "Slot called!";
}
```

You need to emit the signal after the connection to invoke the slot. In the preceding example, we used the traditional way of signal and slot declaration. You can replace the connection with the latest syntax, as shown here:

```
connect(this, &MyClass::signalName, this,
        &MyClass::slotName);
```

It is not only possible to connect one signal to one slot, but also to connect many slots and signals. Similarly, many signals can be connected to one slot. We will learn how to do that in the next section.

Connecting a single signal to multiple slots

You can connect the same signal to multiple slots. These slots will be called in the same order as the connections are made. Let's consider that a signal named signalX() is connected to three slots called slotA(), slotB(), and slotC(). When signalA() is emitted, all three slots will be invoked.

Let's look at the traditional way of making connections:

```
QObject::connect(this, SIGNAL(signalX()),this,
                SLOT(slotA()));
QObject::connect(this, SIGNAL(signalX()),this,
                SLOT(slotB()));
QObject::connect(this, SIGNAL(signalX()),this,
                SLOT(slotC()));
```

You can also create connections as per the new syntax, as follows:

```
connect(this, &MyClass:: signalX, this, &MyClass:: slotA);
connect(this, &MyClass:: signalX, this, &MyClass:: slotB);
connect(this, &MyClass:: signalX, this, &MyClass:: slotC);
```

In the next section, we will learn how to connect multiple signals to a single slot.

Connecting multiple signals to a single slot

In the previous section, you learned how to create a connection between a single signal and multiple slots. Now, let's look at the following code to understand how to connect multiple signals to a single slot:

```
QObject::connect(this, SIGNAL(signalX()),this,
            SLOT(slotX()));
QObject::connect(this, SIGNAL(signalY()),this,
            SLOT(slotX()));
QObject::connect(this, SIGNAL(signalZ()),this,
            SLOT(slotX()));
```

Here, we have used three different signals called `signalX()`, `signalY()`, and `signalZ()`, but there is a single slot defined as `slotX()`. When any of these signals are emitted, that slot is called.

In the next section, we will learn how to connect one signal to another signal.

Connecting a signal to another signal

Sometimes, you may have to forward a signal instead of directly connecting to a slot. You can connect one signal to another signal as follows:

```
connect(sender, SIGNAL(signalA()),forwarder,
       SIGNAL(signalB())));
```

You can also create connections as per the new syntax, as follows:

```
connect(sender,&ClassName::signalA,forwarder,&ClassName::
       signalB);
```

In the preceding line, we have connected `signalA()` to `signalB()`. Hence, when `signalA()` is emitted, `signalB()` will also be emitted and the corresponding slot connected to `signalB()` will be invoked. Let's consider that we have a button in our GUI, and we want to forward the button click as a different signal. The following code snippet shows how to forward a signal:

```
#include <QWidget>
class QPushButton;
class MyClass : public QWidget
{
```

```
    Q_OBJECT
public:
    MyClass(QWidget *parent = nullptr);
    ~MyClass();
signals:
    void signalName();
 private:
    QPushButton *myButton;
};
MyClass::MyClass(QWidget *parent)
    : QWidget(parent)
{
    myButton = new QPushButton(this);
    connect(myButton, &QPushButton::clicked,
            this, &MyClass::signalName);
}
```

In the preceding example, we forwarded the button click signal to our custom signal. We can call the slot that is connected to the custom signal as discussed earlier.

In this section, we learned how connections are made and how to use signals and slots. Now, you can communicate between different classes and share information. In the next section, we will learn about the working mechanism behind signals and slots.

The working mechanism of Qt signals and slots

In the previous sections, we learned about signal and slot syntaxes and how to connect them. Now, we will understand how it works.

While creating a connection, Qt looks for the index of the signal and the slot. Qt uses a lookup string table to find the corresponding indexes. Then, a QObjectPrivate::Connection object is created and added to the internal linked lists. Since one signal can be connected to multiple slots, each signal can have a list of the connected slots. Each connection contains the receiver's name and the index of the slot. Each object has a connection vector that associates with each signal in a linked list of QObjectPrivate::Connection.

The following diagram illustrates how `ConnectionList` creates connections between sender and receiver objects:

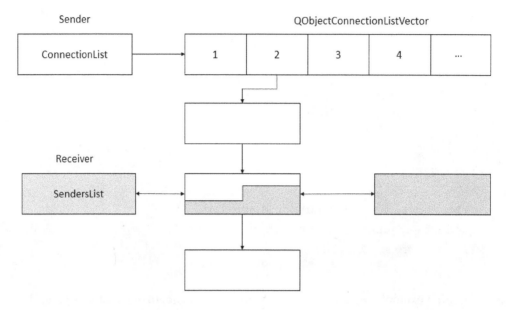

Figure 6.1 – Illustration of the connection mechanism between the sender and receiver

`ConnectionList` is a singly linked list that contains all the connections from and to an object. `signalVector` contains the lists of connections for a given signal. Each `Connection` is also part of a *senders* linked list. Linked lists are used because they permit faster addition and removal of objects. Each object also has a reversed list of connections the object is connected to for automatic deletion. For detailed internal implementation, look at the latest `qobject_p.h`.

There are quite a few articles on how signals and slots work on the *woboq* website. You can also explore the Qt source code on the woboq website. Go to the following link if you need further information:

`https://woboq.com/blog/how-qt-signals-slots-work.html`.

Now, let's learn about Qt's meta-object system.

Qt's meta-object system

Qt's meta-object system is the core mechanism behind the signals and slots mechanism. It provides features such as inter-object communication, a dynamic property system, and runtime type information.

The meta-object system is implemented with a three-part mechanism. These mechanisms are as follows:

- QObject
- Q_OBJECT macro
- Meta-Object Compiler

The QObject class is the base class of all Qt objects. It is a very powerful mechanism that facilitates the signals and slots mechanism. The QObject class provides a base class for objects that can take advantage of the meta-object system. QObject derived classes arrange themselves in an object tree. This creates the parent-children relationship between classes. When you create a QObject derived class with another QObject derived class as a parent, the object will be automatically added to the parent's children() list. The parent takes ownership of the object. GUI programming requires both runtime efficiency and a high level of flexibility. Qt achieved this by combining the speed of C++ with the flexibility of the Qt Object Model. Qt provides the required features by implementing standard C++ techniques based on the inheritance from QObject.

You can learn more about the Qt Object Model at the following link:

https://doc.qt.io/qt-6/object.html.

The Q_OBJECT macro appears inside the private section of the class declaration. It is used to enable signals, slots, and other services provided by Qt's meta-object system.

Meta-Object Compiler (**moc**) generates additional code for QObject derived classes to implement meta-object features. It provides the capability to introspect the objects at runtime. By default, C++ does not have introspection support. Hence, Qt created moc. It is a code-generating program that handles Qt's C++ extensions. The tool reads the C++ header files and if it locates the Q_OBJECT macro, then it creates another C++ source file with the meta-object code. That generated file contains the code required for the introspection. Both files are compiled and linked together. In addition to providing the signals and slots mechanism for communication between objects, the meta-object code offers several additional features to find the class name and inheritance details, and also helps with setting properties at runtime. Qt's moc provides a clean way to go beyond the compiled language's facilities.

You can perform type casts by using `qobject_cast()` on `QObject` derived classes. The `qobject_cast()` function acts similar to the standard C++ `dynamic_cast()`. The advantage is that it doesn't require **runtime type identification (RTTI)** and it works across dynamic library boundaries. You can derive a class from `QObject`, but if you don't add the `Q_OBJECT` macro, then the signals and slots and the other meta-object system features will not be available. A `QObject` derived class without meta code is equivalent to its closest ancestor containing meta-object code. There is also a lighter version of the `Q_OBJECT` macro, known as `Q_GADGET`, that can be used to utilize some of the capabilities provided by `QMetaObject`. A class that uses `Q_GADGET` doesn't have signals or slots.

We have seen a few new keywords here, such as `Q_OBJECT`, `signals`, `slots`, `emit`, `SIGNAL`, and `SLOT`. These are known as the Qt extensions of C++. They are very simple macros meant to be seen by `moc`, defined in `qobjectdefs.h`. Out of these, `emit` is an empty macro that is not parsed by `moc`. It is kept just to give the developer hints.

You can learn about why Qt uses `moc` for signals and slots at `https://doc.qt.io/qt-6/why-moc.html`.

In this section, we learned about Qt's meta-object system. In the next section, we will discuss the `moc` generated code and discuss some of the underlying implementations.

MOC generated code

In this section, we will have a look at the code generated by `moc` in Qt6. When you build the earlier signal and slot example, you will see the generated files under the build directory: `moc_myclass.cpp` and `moc_predefs.h`. Let's open the `moc_myclass.cpp` file with a text editor:

```
#include <memory>
#include "../../SignalSlotDemo/myclass.h"
#include <QtCore/qbytearray.h>
#include <QtCore/qmetatype.h>
#if !defined(Q_MOC_OUTPUT_REVISION)
#error "The header file 'myclass.h' doesn't include
        <QObject>."
#elif Q_MOC_OUTPUT_REVISION != 68
#error "This file was generated using the moc from 6.0.2.
        It"
```

```
#error "cannot be used with the include files from this
        version of Qt."
#error "(The moc has changed too much.)"
#endif
```

You can see that the information about the Qt Meta-Object Compiler version at the top of the file. Please note that all the changes that are made in this file will be lost on recompiling the project. So, don't modify anything in this file. We are looking at the file to understand the working mechanism.

Let's look at the integer data of QMetaObject. As you can see, there are two columns; the first column is the count, while the second column is the index in this array:

```
static const uint qt_meta_data_MyClass[] = {
 // content:
        9,          // revision
        0,          // classname
        0,     0,   // classinfo
        2,    14,   // methods
        0,     0,   // properties
        0,     0,   // enums/sets
        0,     0,   // constructors
        0,          // flags
        1,          // signalCount
 // signals: name, argc, parameters, tag, flags, initial
 // metatype offsets
        1,     0,    26,     2, 0x06,     0 /* Public */,
 // slots: name, argc, parameters, tag, flags, initial
 // metatype offsets
        3,     0,    27,     2, 0x0a,     1 /* Public */,
 // signals: parameters
    QMetaType::Void,
 // slots: parameters
    QMetaType::Void,
        0           // eod
};
```

In this case, we have one method, and the description of the method starts at index 14. You can find the number of available signals in `signalCount`. For each function, moc also saves the return type of each parameter, their type, and their index to the name. In each meta-object, the methods are given an index, beginning with 0. They are arranged as signals, then slots, and then as other functions. These indexes are relative indexes and exclude the indexes of parent objects.

When you look further into the code, you will find the `MyClass::metaObject()` function. This function returns `QObject::d_ptr->dynamicMetaObject()` for dynamic meta-objects. The `metaObject()` function normally returns the class' `staticMetaObject`:

```
const QMetaObject *MyClass::metaObject() const
{
    return QObject::d_ptr->metaObject
? QObject::d_ptr->dynamicMetaObject()
: &staticMetaObject;
}
```

When the incoming string data matches the current class, you must convert this pointer into a void pointer and pass it to the outside world. If it is not the current class, then `qt_metacast()` of the parent class is called to continue the query:

```
void *MyClass::qt_metacast(const char *_clname)
{
    if (!_clname) return nullptr;
    if (!strcmp(_clname,
                qt_meta_stringdata_MyClass.stringdata0))
        return static_cast<void*>(this);
    return QObject::qt_metacast(_clname);
}
```

Qt's meta-object system uses the `qt_metacall()` function to access the meta-information for a particular `QObject` object. When we emit a signal, `qt_metacall()` is called and then calls the real signal function:

```
int MyClass::qt_metacall(QMetaObject::Call _c, int _id, void **_a)
{
    _id = QObject::qt_metacall(_c, _id, _a);
```

```
    if (_id < 0)
        return _id;
    if (_c == QMetaObject::InvokeMetaMethod) {
        if (_id < 2)
            qt_static_metacall(this, _c, _id, _a);
        _id -= 2;
    } else if (_c == QMetaObject::
                    RegisterMethodArgumentMetaType) {
        if (_id < 2)
            *reinterpret_cast<QMetaType *>(_a[0]) =
                                        QMetaType();
        _id -= 2;
    }
    return _id;
}
```

When you call a signal, it calls the moc generated code, which internally calls QMetaObject::activate(), as shown in the following snippet. Then, QMetaObject::activate() looks into the internal data structures to find out about the slots that are connected to that signal.

You can find the detailed implementation of this function inside qobject.cpp:

```
void MyClass::signalName()
{
    QMetaObject::activate(this, &staticMetaObject, 0,
                        nullptr);
}
```

By doing this, you can explore the complete generated code and look at the symbols further. Now, let's look at the moc generated code where the slot is called. The slot is called by its index in the qt_static_metacall function, as shown here:

```
void MyClass::qt_static_metacall(QObject *_o,
    QMetaObject::Call _c, int _id, void **_a)
{
    if (_c == QMetaObject::InvokeMetaMethod) {
        auto *_t = static_cast<MyClass *>(_o);
        (void)_t;
```

```
        switch (_id) {
        case 0: _t->signalName(); break;
        case 1: _t->slotName(); break;
        default: ;
        }
    } else if (_c == QMetaObject::IndexOfMethod) {
        int *result = reinterpret_cast<int *>(_a[0]);
        {
            using _t = void (MyClass::*)();
            if (*reinterpret_cast<_t *>(_a[1]) ==
                static_cast<_t>(&MyClass::signalName)) {
                *result = 0;
                return;
            }
        }
    }
    (void)_a;
}
```

The array pointers to the argument are in the same format as the signal. `_a[0]` is not touched because everything here returns void:

```
bool QObject::isSignalConnected(const QMetaMethod &signal)
const
```

This returns `true` if the signal is connected to at least one receiver; otherwise, it returns `false`.

When an object is destroyed, an `QObjectPrivate::senders` list is iterated, and all `Connection::receiver` are set to 0. Also, `Connection::receiver->connectionLists->dirty` is set to `true`. Each `QObjectPrivate::connectionLists` is also iterated to remove the **connection** in the senders lists.

In this section, we went through some sections of the moc generated code and understood the working mechanism behind signals and slots. In the next section, we will learn about Qt's property system.

Getting to know Qt's property system

Qt's property system is similar to some other compiler vendors. However, it provides a cross-platform advantage and works with standard compilers supported by Qt on different platforms. To add a property, you must add the Q_PROPERTY() macro to the QObject derived class. This property acts like a class data member, but it provides extra features that are available through the Meta-Object System. A simple syntax looks as follows:

```
Q_PROPERTY(type variableName READ getterFunction
           WRITE setterFunction  NOTIFY signalName)
```

In the preceding syntax, we used some of the most common parameters. But there are more parameters that are supported in the syntax. You can find out more by reading the Qt documentation. Let's have a look at the following code snippet, which uses the MEMBER parameter:

```
    Q_PROPERTY(QString text MEMBER m_text NOTIFY
               textChanged)
signals:
    void textChanged(const QString &newText);
private:
    QString m_text;
```

In the preceding snippet, we exported a member variable as a Qt property using the MEMBER keyword. The type here is QString, and the NOTIFY signal is used to implement QML property binding.

Now, let's explore how to read and write properties with the Meta-Object System.

Reading and writing properties with the Meta-Object System

Let's create a class named MyClass, which is a subclass of QWidget. Let's add the Q_OBJECT macro to its private section to enable the property system. In this example, we want to create a property in MyClass to keep track of a version's value. The name of the property will be version, and its type will be QString, which is defined in MyClass. Let's look at the following code snippet:

```
class MyClass : public QWidget
{
    Q_OBJECT
```

```
        Q_PROPERTY(QString version READ version WRITE
                   setVersion NOTIFY versionChanged)
public:
    MyClass(QWidget *parent = nullptr);
    ~MyClass();
    void setVersion(QString version)
    {
        m_version = version;
        emit versionChanged(version);
    }
    QString version() const { return m_version; }
    signals:
        void versionChanged(QString version);
    private:
        QString m_version;
};
```

To get the property changed notification, you have to emit `versionChanged()` after the `version` value is changed.

Let's have a look at the `main.cpp` file for the preceding example:

```
int main(int argc, char *argv[])
{
    QApplication a(argc, argv);
    MyClass myClass;
    myClass.setVersion("v1.0");
    myClass.show();
    return a.exec();
}
```

In the preceding code snippet, the property is set by invoking `setVersion()`. You can see that `versionChanged()` signal is emitted every time the version is changed.

You can also read a property using `QObject::property()` and write it using `QObject::setProperty()`. You can also query dynamic properties using `QObject::property()`, similar to compile time `Q_PROPERTY()` declarations.

You can also set the property like so:

```
QObject *object = &myClass;
object->setProperty("version", "v1.0");
```

In this section, we discussed the property system. In the next section, we will learn about signals and slots in Qt Designer.

Using signals and slots in Qt Designer

If you are using the Qt Widgets module, then you can use Qt Designer to edit signal and slot connections in the form. Qt default widgets come with many signals and slots. Let's see how we can implement signals and slots in Qt Designer without writing any code.

You can drag a **Dial** control and a **Slider** control onto the form. You can add connections via **Signals and Slots Editor** at the bottom tab, as shown in the following screenshot:

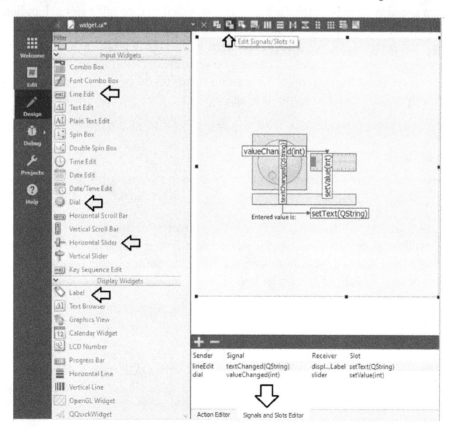

Figure 6.2 – Creating signals and slots connections using Qt Designer

Alternatively, you can press *F4* or select the **Edit Signals/Slots** button from the top toolbar. Then, you can select the control and create a connection by dragging the connection to the receiver. If you have custom signals or slots defined for your custom class, they will be automatically displayed in **Signals and Slots Editor**. However, most developers prefer to define connections inside C++ source files.

In this section, we discussed using Qt Designer to implement signals and slots in Qt Widgets. Now, let's look at how signals are handled in QML.

Understanding signals and the handler event system in QML

Previously, we learned how to connect signals and slots inside C++ source files and use them with the Qt Widgets module. Now, let's look at how we can communicate in QML. QML has a signal and handler mechanism, similar to signals and slots. In a QML document, a signal is an event, and the signal is responded to through a signal handler. Like a slot in C++, a signal handler is invoked when a signal is emitted in QML. In Qt terminology, the method is a slot that is connected to the signal; all the methods defined in QML are created as Qt slots. Hence, there is no separate declaration for slots in QML. A signal is a notification from an object that some event has occurred. You can place logic inside JavaScript or a method to respond to the signal.

Let's look at how to write a signal handler. You can declare a signal handler as follows:

```
onSignalName : {
//Logic
}
```

Here, `signalName` is the name of the signal. The first letter of the signal's name is capitalized while writing a handler. So, the signal handler here is named `onSignalName`. The signal and signal handler should be defined inside the same object. The logic inside the signal handler is a block of JavaScript code.

For example, when the user clicks within the mouse area, the `clicked()` signal is emitted. To handle the `clicked()` signal, we must add the `onClicked: {...}` signal handler.

Signal handlers are simple functions that are invoked by the QML engine when an associated signal is emitted. When you add a signal to a QML object, Qt automatically adds a corresponding signal handler to the object definition.

Let's start by adding a custom signal to a QML document.

Adding a signal in QML

To add a signal inside your QML class, you must use the `signal` keyword. The syntax for defining a new signal is as follows:

```
signal <name>[(([<type> <parameter name>[...]])]
```

The following is an example of this:

```
signal composeMessage(string message)
```

A signal can be declared with or without any parameters. If no parameter is declared for the signal, then you can leave () brackets. You can emit a signal by invoking it as a function:

```
Rectangle {
    id: mailBox
    signal composeMessage(string message)
    anchors.fill: parent
    Button {
        id:sendButton
        anchors.centerIn: parent
        width: 100
        height: 50
        text: "Send"
        onClicked:   mailBox.composeMessage("Hello World!")
    }
    onComposeMessage: {
        console.log("Message Received",message)
    }
}
```

In the preceding example, we added a custom signal composeMessage() to the QML file. We used the corresponding signal handler; that is, onComposeMessage(). Then, we added a button that emits the composeMessage() signal when it is clicked. When you run this example, you will see that the signal handler is called automatically when the button is clicked.

In this section, you learned how to declare a signal and how to implement the corresponding signal handler. In the next section, we will connect the signal to a function.

Connecting a signal to a function

You can connect a signal to any function defined inside your QML document. You can use connect () to connect a signal either to a function or another signal. When a signal is connected to a function, that function is automatically invoked whenever the signal is emitted. This mechanism enables a signal to be received by a function instead of a signal handler.

In the following snippet, the composeMessage () signal is connected to the transmitMessage () function using the connect () function:

```qml
Rectangle {
    id: mailBox
    signal composeMessage(string message)
    anchors.fill: parent
    Text {
        id: textElement
        anchors {
            top:   parent.top
            left: parent.left
            right:parent.right
        }
        width: 100
        height:50
        text: ""
        horizontalAlignment: Text.AlignHCenter
    }
    Component.onCompleted: {
        mailBox.composeMessage.connect(transmitMessage)
        mailBox.composeMessage("Hello World!")
    }
    function transmitMessage(message) {
        console.log("Received message: " + message)
        textElement.text = message
    }
}
```

In QML, signal handling is implemented using the following syntax:

```
sender.signalName.connect(receiver.slotName)
```

You can also remove a connection using the disconnect() function. You can disconnect the connection like so:

```
sender.signalName.disconnect(receiver.slotName)
```

Now, let's explore how to forward a signal in QML.

Connecting a signal to a signal

You can connect a signal to another signal in QML. You can achieve this using the connect() function.

Let's explore how we can do this by looking at the following example:

```
Rectangle {
    id: mailBox
    signal forwardButtonClick()
    anchors.fill: parent
    Button {
        id:sendButton
        anchors.centerIn: parent
        width: 100
        height: 50
        text: "Send"
    }
    onForwardButtonClick: {
        console.log("Fordwarded Button Click Signal!")
    }
    Component.onCompleted: {
        sendButton.clicked.connect(forwardButtonClick)
    }
}
```

In the preceding example, we connected the `clicked()` signal to the `forwardButtonClick()` signal. You can implement the necessary logic at the root level inside the `onForwardButtonClick()` signal handler. You can also emit the signal from the button click handler, as follows:

```
onClicked: {
    mailBox.forwardButtonClick()
}
```

In this section, we discussed how to connect two signals and handle them. In the next section, we will discuss how to communicate between the C++ class and QML using signals and slots.

Defining property attributes and understanding property binding

Previously, we learned how to define a type in C++ by registering the `Q_PROPERTY` of a class, which is then registered with the QML type system. It is also possible to create custom properties in a QML document. Property binding is a core feature of QML that allows us to create relationships between various object properties. You can declare a property in a QML document with the following syntax:

```
[default] property <propertyType> <propertyName> : <value>
```

In this way, you can expose a particular parameter to outside objects or maintain internal states more efficiently. Let's look at the following property declaration:

```
property string version: "v1.0"
```

When you declare a custom property, Qt implicitly creates a property-change signal for that property. The associated signal handler is on<`PropertyName`>Changed, where <`PropertyName`> is the name of the property, with the first letter capitalized. For the previously declared property, the associated signal handler is `onVersionChanged`, as shown here:

```
onVersionChanged:{...}
```

If the property is assigned a static value, then it remains constant until it is explicitly assigned a new value. To update these values dynamically, you should use property binding inside your QML document. We used simple property binding earlier, as shown in the following snippet:

```
width: parent.width
```

However, we can combine this with the property that's exposed by the backend C++ class, as shown here:

```
property string version: myClass.version
```

In the previous line, `myClass` is the backend C++ object that's registered with the QML engine. In this case, whenever the `versionChanged()` change signal is emitted from the C++ side, the QML `version` property gets updated automatically.

Next, we'll discuss how to integrate signals and slots between C++ and QML.

Integrating signals and slots between C++ and QML

In C++, to interact with the QML layer, you can use signals, slots, and `Q_INVOKABLE` functions. You can also create properties using the `Q_PROPERTY` macro. To respond to signals from objects, you can use the `Connections` QML type. When a property changes inside a C++ file, `Q_PROPERTY` automatically updates the values. If the property has a binding with any QML property, it will automatically update the property values inside QML. In this case, the signal slot mechanism is established automatically.

Let's look at the following example, which uses the aforementioned mechanism:

```cpp
class CPPBackend : public QObject
{
    Q_OBJECT
    Q_PROPERTY(int counter READ counter WRITE setCounter
                NOTIFY counterChanged)
public:
    explicit CPPBackend(QObject *parent = nullptr);
    Q_INVOKABLE  void receiveFromQml();
    int counter() const;
    void setCounter(int counter);
signals:
    void sendToQml(int);
    void counterChanged(int counter);
private:
    int m_counter = 0;
};
```

In the preceding code, we declared a Q_PROPERTY-based notification. We can get the new `counter` value when the `counterChanged()` signal is emitted. However, we used the `receiveFromQml()` function as a `Q_INVOKABLE` function so that we can call it directly inside the QML document. We are emitting `sendToQml()`, which is handled inside `main.qml`:

```cpp
void CPPBackend::setCounter(int counter)
{
    if (m_counter == counter)
        return;
    m_counter = counter;
    emit counterChanged(m_counter);
}
void CPPBackend::receiveFromQml()
{
    // We increase the counter and send a signal with new
    // value
    ++m_counter;
    emit sendToQml(m_counter);
}
```

Now, let's have a look at the QML implementation:

```qml
Window {
    width: 640
    height: 480
    visible: true
    title: qsTr("C++ QML Signals & Slots Demo")
    property int count: cppBackend.counter
    onCountChanged:{
        console.log("property is notified. Updated value
                    is:",count)
    }
    Connections {
        target: cppBackend
        onSendToQml: {
            labelCount.text ="Fetched value is "
                            +cppBackend.counter
```

```
                }
        }
        Row{
                anchors.centerIn: parent
                spacing: 20
                Text {
                        id: labelCount
                        text: "Fetched value is " + cppBackend.counter
                }
                Button {
                        text: qsTr("Fetch")
                        width: 100 ;height: 20
                        onClicked: {
                                cppBackend.receiveFromQml()
                        }
                }
        }
}
```

In the preceding example, we used Connections to connect to the C++ signal. On button click, we are calling the receiveFromQml() C++ function, where we are emitting the signal. We have also declared the count property, which also listens to counterChanged(). We handled the data inside the associated signal handler; that is, onCountChanged. We can also update the labelCount data based on the notification:

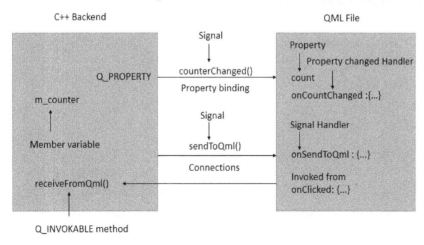

Figure 6.3 – The mechanism that was used in this example

The preceding diagram illustrates the communication mechanism in this example. For explanation purposes, we have kept multiple approaches in the same example to explain the communication mechanism between C++ and QML.

In this section, you learned about the signals and slots mechanism through examples. In the next section, we will learn about events and the event loop in Qt.

Understanding events and the event loop

Qt is an event-based system, and all GUI applications are event-driven. In an event-driven application, there is usually a main loop that listens for events and then triggers a callback function when one of those events is detected. Events can be spontaneous or synthetic. Spontaneous events come from the outside environment. Synthetic events are custom events generated by the application. An event in Qt is a notification that represents something that has happened. Qt events are value types, derived from QEvent, which offers a type enumeration for each event. All events that arise inside a Qt application are encapsulated in objects that inherit from the QEvent class. All QObject derived classes can override the QObject::event() function in order to handle events targeted by their instances. Events can come from both inside and outside the application.

When an event occurs, Qt produces an event object by constructing an appropriate QEvent subclass instance, which it then delivers to the specific instance of QObject by calling its event() function. Unlike the signals and slots mechanism, where the slots connected to the signal are usually executed immediately, an event must wait for its turn, until the event loop dispatches all the events that arrived earlier. You must select the right mechanism as per your intended implementation. The following diagram illustrates how events are created and managed in event-driven applications:

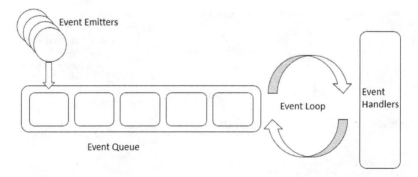

Figure 6.4 – Illustration of an event-driven application using the event loop

We can enter Qt's main event loop by calling QCoreApplication::exec().
The application keeps running until QCoreApplication::exit() or
QCoreApplication::quit() are called, which terminates the loop.
QCoreApplication can process each event in the GUI thread and forward events
to QObjects. Please note that the events are not delivered immediately; instead, they're
queued up in an event queue and processed later, one after another. The event dispatcher
loops through this queue, converts them into QEvent objects, and then dispatches the
events to the target QObject.

A simplified event loop dispatcher may look as follows:

```
while(true)
{
    dispatchEventsFromQueue();
    waitForEvents();
}
```

Some of the important Qt classes related to the event loop are as follows:

- **QAbstractEventDispatcher** is subclassed to manage Qt's event queue.
- **QEventLoop** provides a local event loop.
- **QCoreApplication** provides an event loop for non-GUI based applications.
- **QGuiApplication** contains the main event loop for GUI-based applications.
- **QThread** is used to create custom threads and manage threads.
- **QSocketNotifier** is used to monitor activity on a file descriptor.
- **QTimer** is used to create a timer in any thread with an event loop.

You can read about these classes in the Qt documentation. The following link provides
further insight into the event system:

https://wiki.qt.io/Threads_Events_QObjects.

In this section, we discussed events and Qt's event loop. In the next section, we will learn
how to filter events with an event filter.

Managing events with an event filter

In this section, you will learn how to manage events and how to filter a specific event and perform a task. You can achieve event filtering by reimplementing event handlers and installing event filters. You can redefine what an event handler should do by subclassing the widget of interest and reimplementing that event handler.

Qt provides five different approaches for event processing, as follows:

- Reimplementing a specific event handler, such as paintEvent()
- Reimplementing the QObject::event() function
- Installing an event filter on the QObject instance
- Installing an event filter on the QApplication instance
- Subclassing QApplication and reimplementing notify()

The following code handles the left mouse button click on a custom widget while passing all other button clicks to the base QWidget class:

```
void MyClass::mousePressEvent(QMouseEvent *event)
{
    if (event->button() == Qt::LeftButton)
    {
        // Handle left mouse button here
    }
    else
    {
        QWidget::mousePressEvent(event);
    }
}
```

In the previous example, we filtered only the left button press event. You can add the required action inside the respective block. The following diagram illustrates the high-level event processing mechanism:

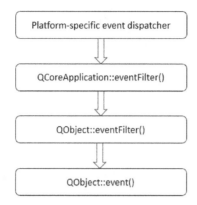

Figure 6.5 – Illustration of the event filter mechanism

An event filter can be installed either on an application instance or a local object. If the event filter is installed in an QCoreApplication object, then all the events will go through this event filter. If it is installed in a QObject derived class, then the events that have been sent to that object will go through the event filter. Sometimes, there may not be any suitable Qt event type available for a specific action. In that case, you can create a custom event by creating a subclass from QEvent. You can reimplement QObject::event() to filter the intended event, as shown here:

```cpp
#include <QWidget>
#include <QEvent>
class MyCustomEvent : public QEvent
{
public:
    static const QEvent::Type MyEvent
                = QEvent::Type(QEvent::User + 1);
};
class MyClass : public QWidget
{
    Q_OBJECT
public:
    MyClass(QWidget *parent = nullptr);
    ~MyClass();
protected:
    bool event(QEvent *event);
};
```

Here, we have created a custom event class named MyCustomEvent and created a custom type.

Now, let's filter these events by reimplementing event():

```cpp
bool MyClass::event(QEvent *event)
{
    if (event->type() == QEvent::KeyPress)
    {
        QKeyEvent *keyEvent= static_cast<QKeyEvent
                                             *>(event);
        if (keyEvent->key() == Qt::Key_Enter)
        {
            // Handle Enter event event
            return true;
        }
    }
    else if (event->type() == MyCustomEvent::MyEvent)
    {
        MyCustomEvent *myEvent = static_cast<MyCustomEvent
                                *>(event);
        // Handle custom event
        return true;
    }
    return QWidget::event(event);
}
```

As you can see, we have passed the other events to QWidget::event() for further processing. If you want to prevent the event from propagating further, then return true; otherwise, return false.

An event filter is an object that receives all the events that are sent to an object. The filter can either stop the event or forward it to the object. It screens the events if an object has been installed as an event filter for the watched object. It is also possible to monitor another object for events by using an event filter and performing the necessary tasks. The following example shows how to reimplement one of the most commonly used events – a keypress event – using the *event filter* approach.

Let's have a look at the following code snippet:

```cpp
#include <QMainWindow>
class QTextEdit;
class MainWindow : public QMainWindow
{
    Q_OBJECT
public:
    MainWindow(QWidget *parent = nullptr);
    ~MainWindow();
protected:
    bool eventFilter(QObject *obj, QEvent *event) override;
private:
    QTextEdit *textEdit;
};
```

In the preceding code, we created a class called `MainWindow` and overridden `eventFilter()`. Let's install the filter on `textEdit` using `installEventFilter()`. You can install multiple event filters on an object. However, if multiple event filters are installed on a single object, the filter that was installed last will be activated first. You can also remove an event filter by calling `removeEventFilter()`:

```cpp
#include "mainwindow.h"
#include <QTextEdit>
#include <QKeyEvent>
MainWindow::MainWindow(QWidget *parent)
    : QMainWindow(parent)
{
    textEdit = new QTextEdit;
    setCentralWidget(textEdit);
    textEdit->installEventFilter(this);
}
```

In the preceding code, we installed an eventFilter on the textEdit object. Now, let's have a look at the eventFilter() function:

```cpp
bool MainWindow::eventFilter(QObject *monitoredObj, QEvent
*event)
{
    if (monitoredObj == textEdit)
    {
        if (event->type() == QEvent::KeyPress)
        {
            QKeyEvent *keyEvent = static_cast<QKeyEvent*>
                                        (event);
            qDebug() << "Key Press detected: " <<
                                        keyEvent->text();
            return true;
        }
        else
        {
            return false;
        }
    }
    else
    {
        return QMainWindow::eventFilter(monitoredObj,
                                        event);
    }
}
```

Here, textEdit is the monitored object. Every time you press a key and if the textEdit is in focus, the event is captured. Since there may more objects that are children and QMainWindow may require the events, don't forget to pass the unhandled events to the base class for further event processing.

> **Important note**
>
> Once you've consumed the event in the eventFilter() function, make sure to return true. If the receiver object is deleted and you return false, then it can result in an application crash.

You can also combine the signals and slots mechanism with the event. You can achieve this by filtering the event and emitting a signal that corresponds to that event. I hope that you have understood the event handling mechanism in Qt. Now, let's look at drag and drop.

Drag and drop

In this section, we will learn about **drag and drop (DnD)**. In a GUI application, DnD is a pointing device gesture in which the user selects a virtual object by *grabbing* it and then *releasing* it on another virtual object. The drag and drop operation starts when the user makes some gesture that is recognized as a signal to start a drag action.

Let's discuss how we can implement drag and drop using Qt Widgets.

Drag and drop in Qt Widgets

In Qt Widgets-based GUI applications, where drag and drop is used, the user starts dragging from a particular widget and drops the dragged object onto another widget. This requires us to reimplement several functions and it handles the corresponding events. The most common functions that need to be reimplemented to achieve drag and drop are as follows:

```
void dragEnterEvent(QDragEnterEvent *event) override;
void dragMoveEvent(QDragMoveEvent *event) override;
void dropEvent(QDropEvent *event) override;
void mousePressEvent(QMouseEvent *event) override;
```

Once you've reimplemented the preceding functions, enable dropping on the target widget with the following statement:

```
setAcceptDrops(true);
```

To begin a drag, create a `QDrag` object and pass a pointer to the widget that begins the drag. The drag and drop operation is handled by a `QDrag` object. This operation requires the attached data description to be of the **Multipurpose Internet Mail Extensions (MIME)** type:

```
QMimeData *mimeData = new QMimeData;
mimeData->setData("text/csv", csvData);
QDrag *dragObject = new QDrag(event->widget());
dragObject->setMimeData(mimeData);
dragObject->exec();
```

The preceding code shows how to create a drag object and set a custom MIME type. Here, we used `text/csv` as the MIME type. You can supply more than one type of MIME-encoded data with a drag and drop operation.

To intercept drag and drop events, you can reimplement `dragEnterEvent()`. This event handler is called when a drag is in progress and the mouse enters the widget.

You can find several relevant examples in the examples section in Qt Creator. Since Qt Widgets aren't very popular these days, we are skipping the drag and drop example using widgets. In the next section, we will discuss drag and drop in QML.

Drag and drop in QML

In the previous section, we discussed drag and drop using widgets. Since QML is used to create modern and touch-friendly applications, drag and drop is a very important feature. Qt provides several convenient QML types for implementing drag and drop. Internally, the corresponding events are handled similarly. These functions are declared in the `QQuickItem` class.

For example, `dragEnterEvent()` is also available in `QQuickItem`, and is used to intercept drag and drop events, as described here:

```
void QQuickItem::dragEnterEvent(QDragEnterEvent *event)
```

Let's discuss how to implement this using the available QML types. Using the `Drag` attached property, any `Item` can be made a source of drag and drop events within a QML scene. A `DropArea` is an invisible item that can receive events when an item is dragged over it. When a drag action is active on an item, any change that's made to its position will generate a drag event that will be sent to any intersecting `DropArea`. The `DragEvent` QML type provides information about a drag event.

The following code snippet shows a simple drag and drop operation in QML:

```
Rectangle {
    id: dragItem
    property point beginDrag
    property bool caught: false
    x: 125; y: 275
    z: mouseArea.drag.active || mouseArea.pressed ? 2 : 1
    width: 50; height: 50
    color: "red"
    Drag.active: mouseArea.drag.active
```

```
Drag.hotSpot.x: 10 ; Drag.hotSpot.y: 10
MouseArea {
id: mouseArea
anchors.fill: parent
drag.target: parent
onPressed: dragItem.beginDrag = Qt.point(dragItem.x,
                                         dragItem.y)
onReleased: {
    if(!dragItem.caught) {
    dragItem.x = dragItem.beginDrag.x
    dragItem.y = dragItem.beginDrag.y
  }
 }
 }
}
```

In the preceding code, we created a draggable item with an ID of `dragItem`. It contains a `MouseArea` to capture the mouse press event. Dragging is not limited to mouse drags. A drag action can be triggered by anything that can generate a drag event. A drag can be canceled by calling `Drag.cancel()` or by setting the `Drag.active` state to `false`.

The drop action can be completed with a drop event by calling `Drag.drop()`. Let's add a `DropArea`:

```
Rectangle {
    x: parent.width/2
    width: parent.width/2 ; height:parent.height
    color: "lightblue"
    DropArea {
    anchors.fill: parent
    onEntered: drag.source.caught = true
    onExited: drag.source.caught = false
   }
}
```

In the preceding snippet, we used a light blue rectangle to distinguish it as a `DropArea` on the screen. We are catching `dragItem` when it enters the `DropArea` region. When `dragItem` is leaving the `DropArea` region, the drop action is disabled. Therefore, when the drop is unsuccessful, the item will go back to its original position.

In this section, we learned about drag and drop actions and their corresponding events. We discussed how to implement them using the Qt Widgets module, as well as in QML. Now, let's summarize what we learned in this chapter.

Summary

In this chapter, we looked at the core concepts of signals and slots in Qt. We discussed different ways of connecting signals and slots. We also learned how to connect one signal to multiple slots and multiple signals to a single slot. Then, we looked at how to use them with Qt Widgets, as well as in QML, as well as the mechanism behind the signal and slot connection. After that, you learned how to communicate between C++ and QML using signals and slots.

This chapter also discussed events and event loops in Qt. We explored how to use events instead of the signal-slot mechanism. After doing this, we created a sample program with a custom event handler to capture events and filter them.

After learning about events, we implemented a simple drag and drop example. Now, you can communicate between classes, between C++ and QML, and implement the necessary actions based on events.

In the next chapter, we will learn about Model View programming and how to create custom models.

7
Model View Programming

Model/View programming is used to separate data from Views in Qt to handle datasets. The **Model/View (M/V)** architecture differentiates the functionalities that give freedom to the developers to modify and present the information on the **User Interface (UI)** in multiple ways. We will discuss each component of the architecture, the related convenience classes offered by Qt, and how to use them with practical examples. Throughout this chapter, we will be discussing the Model View pattern and understand the underlying core concepts.

In this chapter, we will discuss the following topics:

- Fundamentals of the M/V architecture
- Using Models and Views
- Creating custom Models and delegates
- Displaying information using M/V in Qt Widgets
- Displaying information using M/V in QML
- Using C++ Models with QML

By the end of this chapter, you will be able to create a data model and display information on a customized UI. You will be able to write your custom models and delegates. You will also learn to represent the information in a UI through Qt Widgets and QML.

Technical requirements

The technical requirements for this chapter include the minimum versions of Qt 6.0.0 and Qt Creator 4.14.0 installed on one of the latest desktop platforms, such as Windows 10, Ubuntu 20.04, or macOS 10.14.

All the code used in this chapter can be downloaded from the following GitHub link: `https://github.com/PacktPublishing/Cross-Platform-Development-with-Qt-6-and-Modern-Cpp/tree/master/Chapter07`.

> **Important note**
> The screenshots used in this chapter are taken on the Windows platform.
> You will see similar screens based on the underlying platform on your machine.

Understanding the M/V architecture

Traditionally, the **Model-View-Controller** (**MVC**) design pattern is often used when building UIs. As the name suggests, it consists of three terms: Model, View, and Controller. The **Model** is an independent component with a dynamic data structure and logic, the **View** is the visual element, and the **Controller** decides how the UI responds to the user inputs. Before MVC came into existence, developers used to put these components together. It is not always possible to decouple the Controller from other components although developers want to keep them as distinct from each other as possible. MVC design decouples the components to increase flexibility and reuse. The following figure illustrates the components of a traditional MVC pattern:

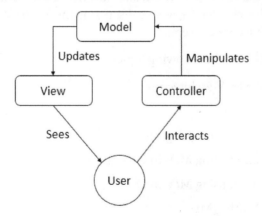

Figure 7.1 – Traditional MVC design pattern

In the MVC pattern, a user sees the View and interacts with a Controller. The Controller sends data to the Model and the Model updates the View. If the View and the Controller components are combined, then it results in the M/V architecture. It provides a more versatile architecture. It is based on the same principle but makes the implementation much simpler. The modified architecture allows us to display the same data in several different Views. The developer can implement new types of Views without changing the underlying data structures. To bring this flexibility to our handling of the user inputs, Qt introduced the concept of **delegate**. Instead of having a Controller, the View receives the data that is updated via a delegate. It has two primary purposes:

- To help the View render each value

- To help the View when the user wants to make some changes

So, in a certain way, the Controller has combined with the View and the View also performs some of the Controller's work through the delegate. The benefit of having a delegate is that it provides the means by which data elements are rendered and modified.

Let's understand the M/V implementation and its components with a diagram:

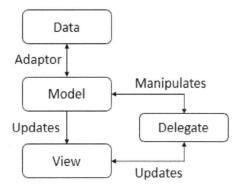

Figure 7.2 – The Qt Model-View-Delegate Framework

As illustrated in *Figure 7.2*, the M/V components are separated into the three sections of **Model**, **View**, and **Delegate**. The **Model** interacts with a database and works as an interface for the architecture's other components. The purpose of the communication is determined by the data source and the model's implementation. The **View** attains the references to items of data known as a **model index**. The View can retrieve the individual item data from the data model by using this model index. In standard Views, a delegate renders the items of data. When an item is modified, the **Delegate** notifies the Model by using the model index.

Figure 7.3 illustrates how a Model delivers data to the View, which is displayed on the individual delegates:

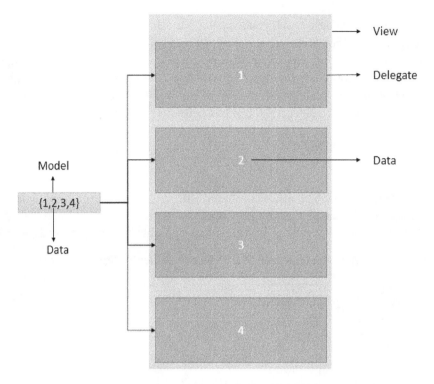

Figure 7.3 – Model-View-Delegate implementation illustration

The Qt framework provides a set of standard classes that implement the M/V architecture to manage the relationship between data and how the user sees it. By decoupling the functionalities, the architecture provides flexibility to customize the presentation of data and allows the combining of an extensive range of data sources with the Views.

The Model, View, and Delegate use a **signal and slot mechanism** to communicate with each other. The Model emits a signal to notify about the data change that occurred in the data source. When a user interacts with the View, then a signal from the View is emitted to notify about the user action. The Delegate emits a signal to notify the Model and View about the edited state.

Now, you have learned the fundamentals of the M/V architecture. The following sections explain how to use the M/V pattern in Qt. We will start with the standard classes provided by the Qt framework and then we will discuss the use of M/V in Qt Widgets. You will learn how to create new components as per the M/V architecture. Let's go ahead!

Model

M/V removes the data consistency challenges that may happen with the standard widgets. It makes it easier to use more than one View for the same data, as one Model can be passed to multiple Views. Qt provides several abstract classes for M/V implementation with common interfaces and certain feature implementations. You can subclass the abstract classes and add the intended functionalities expected by other components. In the M/V implementation, the model provides a standard interface used by the View and delegate to access the data.

Qt offers some ready-made Model classes such as QStandardItemModel, QFileSystemModel, and QSqlTableModel. QAbstractItemModel is the standard interface defined by Qt. The subclasses of QAbstractItemModel represent the data in a hierarchical structure. *Figure 7.4* illustrates the hierarchy of Model classes:

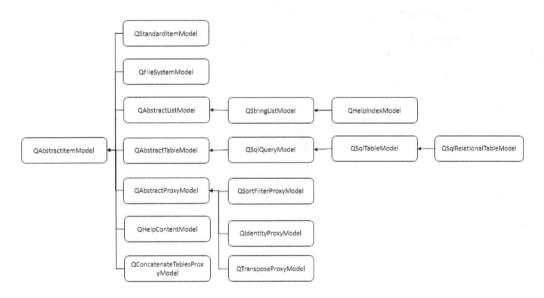

Figure 7.4 – The hierarchy of Model classes in the Qt

Views use this approach to access individual data items in the Model, but they are not restricted in the way that they present this information to the user. The data passed through a Model can be held in a data structure or a database, or some other application component. All item Models are based on the QAbstractItemModel class.

Figure 7.5 shows how data is arranged in different types of Models:

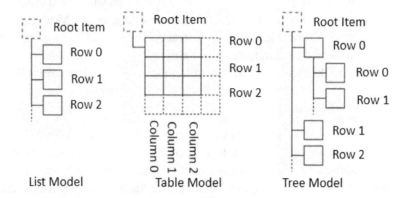

Figure 7.5 – Different types of models and arrangements of data

The data is represented through Models either in a tabular representation in the form of rows and columns, or using a hierarchical representation of the data. In the M/V pattern, widgets do not store data behind the individual cells. They directly use the data. You may have to create a wrapper to make your data compatible with the QAbstractItemModel interface. Views use this interface to read and write the data. Any class that is derived from QAbstractItemModel is known as a Model. It provides an interface to handle Views that represent data in the forms of lists, tables, and trees. To implement a custom Model for a list or a table-like data structure, you can derive from QAbstractListModel and QAbstractTableModel to use the available features. The subclasses provide Models that are suitable for specialized lists and tables.

The Qt framework provides two standard types of Models. They are as follows:

- QStandardItemModel
- QFileSystemModel

The QStandardItemModel is a multi-purpose Model where custom data can be stored. Each element refers to an item. It can be used to display a variety of data structures needed by lists, tables, and tree Views. It provides a traditional item-based approach to dealing with the Model. QStandardItem provides the items used in a QStandardItemModel.

QFileSystemModel is a Model that keeps information about the contents of a directory. It simply represents files and directories on the local filing system and doesn't hold any items of data. It provides a ready-to-use Model to create a sample application and you can manipulate the data using Model indexes. Now, let's discuss what a delegate is.

Delegate

Delegates provide control over the presentation of items displayed in the View. The M/V pattern, unlike the MVC pattern, does not have an entirely different component for handling user interaction. The View is primarily in charge of displaying the Model data to the user and allowing them to interact with it. To add some flexibility to how the user action is obtained, the delegates handle the interactions. It empowers certain widgets to be used as editors for editable items in the Model. Delegates are used to provide interaction capabilities and rendering individual fields in the Views. The `QAbstractItemDelegate` class defines the basic interface for managing delegates. There are a few ready-made delegate classes provided by Qt to use with built-in widgets to modify a particular data type.

To understand it better, we will have a look at the hierarchy of delegate classes in the Qt framework (see *Figure 7.6*):

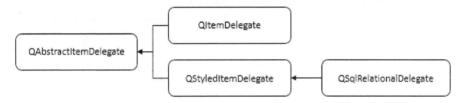

Figure 7.6 – The hierarchy of delegate classes in the Qt framework

As we can see in the preceding diagram, `QAbstractItemDelegate` is the abstract base class for delegates. The default delegate implementation is provided by `QStyledItemDelegate`. Qt's standard Views use it as the default delegate. Other options for painting and creating editors for items in Views are `QStyledItemDelegate` and `QItemDelegate`. You can use `QItemDelegate` to customize display features and editor widgets for an item. The difference between these two classes is that, unlike `QItemDelegate`, `QStyledItemDelegate` uses the current style to paint its items. `QStyledItemDelegate` can handle the most common data types such as `int` and `QString`. It is recommended to subclass `QStyledItemDelegate` while creating new delegates or while working with Qt Style Sheets. By writing a custom delegate, you can use a custom data type or customize the rendering.

In this section, we discussed the different types of Models and delegates. Let's discuss the View classes provided by Qt Widgets.

Views in Qt Widgets

Several convenience classes are derived from the standard View classes to implement the M/V pattern. Examples of such convenience classes are QListWidget, QTableWidget, and QTreeWidget. As per the Qt documentation, these classes are less adaptable than View classes, and they can't be used for random Models. Based on your project requirements, you have to choose suitable widget classes for implementing the M/V pattern.

If you want to use an item-based interface and take advantage of the M/V pattern, then it is recommended to use the following View classes with QStandardItemModel:

- QListView displays a list of items.

- QTableView displays data from a Model in a table.

- QTreeView shows Model items of data in a hierarchical list.

The hierarchy of View classes in the Qt framework is as follows:

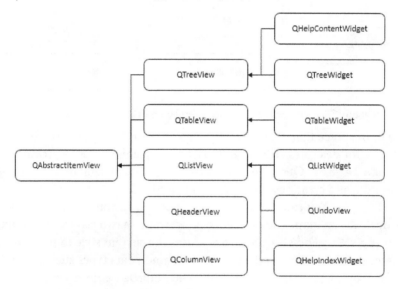

Figure 7.7 – The hierarchy of View classes in the Qt framework

QAbstractItemView is the abstract base class of the aforementioned classes. These classes can be derived to have specialized Views, even though they provide ready-to-use implementations. The most appropriate Views to use for QFileSystemModel are QListView and QTreeView. Each of these Views has its unique way of representing the data items. For example, QTreeView displays a tree hierarchy as a horizontal series of lists. All these Views must have a Model associated with them. There are several predefined Models provided by Qt. You can add your own customized Model if the ready-made Models don't meet your criteria.

Unlike the View classes (for which the class name ends with `View`), the convenience widgets (for which the class name ends with `Widget`) do not need to be backed by a Model and can be used directly. The main advantage of using convenience widgets is that they require the least amount of effort to work with them.

Let's look at the different View classes in the Qt Widgets module and which readymade Models can be used with them:

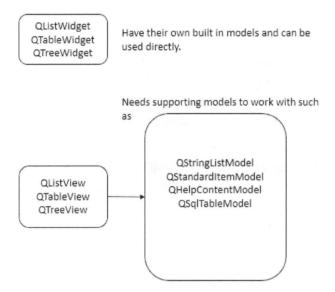

Figure 7.8 – Different types of Qt widgets used as the View in the M/V pattern

The delegate is used to display individual field data in `QListView`, `QTableView`, or `QTreeView`. When a user starts interacting with an item, the delegate provides an editor widget for editing to take place.

You can find a comparative overview of the aforementioned classes and learn about the uses of the corresponding widgets at the following link:

`https://doc.qt.io/qt-6/modelview.html`

In this section, you learned about the M/V architecture and got familiar with the terms used. Let's implement M/V with a simple GUI application using Qt Widgets.

Creating a simple Qt Widgets application using the M/V pattern

It is time for us to create a simple example using *Qt Widgets*. The example in this section illustrates how a predefined `QFileSystemModel` is used in association with the built-in `QListView` and `QTreeView` widgets. Delegation is automatically taken care of when the Views are double-clicked.

Follow these steps to create a simple application that implements the M/V pattern:

1. Create a new project using Qt Creator, selecting the **Qt Widgets** template from the project creation wizard. It will generate a project with a predefined project skeleton.

2. Once the application skeleton is created, open the `.ui` form and add `QListView` and `QTreeView` to the form. You may add two labels to distinguish the Views as shown here:

Figure 7.9 – Create a UI with QListView and QTreeView using Qt Designer

3. Open the `mainwindow.cpp` file and add the following contents:

```cpp
#include "mainwindow.h"
#include "ui_mainwindow.h"
#include <QFileSystemModel>
MainWindow::MainWindow(QWidget *parent)
```

```
        : QMainWindow(parent)
        , ui(new Ui::MainWindow)
{
    ui->setupUi(this);
    QFileSystemModel *model = new QFileSystemModel;
    model->setRootPath(QDir::currentPath());
    ui->treeView->setModel(model);
    ui->treeView->setRootIndex(
        model->index(QDir::currentPath()));
    ui->listView->setModel(model);
    ui->listView->setRootIndex(
        model->index(QDir::currentPath()));
}
```

In the preceding C++ implementation, we have used a predefined
QFileSystemModel as the Model for the Views.

4. Next, hit the **Run** button in the left pane. You will see a window as shown in
    Figure 7.10 once you hit the **Run** button:

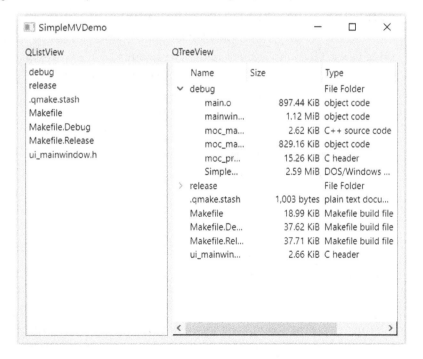

Figure 7.10 – Output of the sample application showing QListView and QTreeView

5. Let's modify the existing application to use a custom Model derived from QAbstractItemModel. In the following example, we have created a simple ContactListModel custom class derived from QAbstractItemModel:

```cpp
void ContactListModel::addContact(QAbstractItemModel
*model,
const QString &name,const QString &phoneno, const QString
&emailid)
{
    model->insertRow(0);
    model->setData(model->index(0, 0), name);
    model->setData(model->index(0, 1), phoneno);
    model->setData(model->index(0, 2), emailid);
}
QAbstractItemModel* ContactListModel::
        getContactListModel()
{
    QStandardItemModel *model = new
        QStandardItemModel(0, 3, this);
    model->setHeaderData(0,Qt::Horizontal,
                        QObject::tr("Name"));
    model->setHeaderData(1,Qt::Horizontal,
                        QObject::tr("Phone No"));
    model->setHeaderData(2,Qt::Horizontal,
                        QObject::tr("Email ID"));

    addContact(model,"John","+1
            1234567890","john@abc.com");
    addContact(model,"Michael","+44
            213243546","michael@abc.com");
    addContact(model,"Robert","+61
            5678912345","robert@xyz.com");
    addContact(model,"Kayla","+91
            9876554321","kayla@xyz.com");
    return model;
}
```

6. Next, modify the UI form to implement a `QTableView` and set the contact list Model to it as shown in the following snippet:

```
ContactListModel *contactModel = new ContactListModel;
ui->tableView->setModel(
               contactModel->getContactListModel());
ui->tableView->horizontalHeader()-
>setStretchLastSection(true);
```

7. You can add `QStringListModel` to the `QListView` to use a simple list Model:

```
QStringListModel *model = new QStringListModel(this);
QStringList List;
List << "Item 1" << "Item 2" << "Item 3" <<"Item 4";
model->setStringList(List);
ui->listView->setModel(model);
```

8. Next, hit the **Run** button in the left pane. You will see a window as shown in *Figure 7.11*:

Figure 7.11 – Output of the application using custom models in QListView and QTableView

Congratulations! You have learned how to use M/V for your Qt widgets project.

> **Important note**
>
> For more implementations of convenience classes such as `QTableWidget` or `QtTreeWidget`, explore the relevant examples on the Qt Creator welcome screen and this chapter's source code.

You can also create your own custom delegate class. To create a custom delegate, you need to subclass `QAbstractItemDelegate` or any of the convenience classes such as `QStyledItemDelegate` or `QItemDelegate`. A custom delegate class may look like the following code snippet:

```cpp
class CustomDelegate: public QStyledItemDelegate
{
  Q_OBJECT
public:
  CustomDelegate(QObject* parent = nullptr);
  void paint(QPainter* painter,
             const QStylestyleOptionViewItem& styleOption,
             const QModelIndex& modelIndex) const override;
  QSize sizeHint(const QStylestyleOptionViewItem& styleOption,
                 const QModelIndex& modelIndex) const override;
  void setModelData(QWidget* editor, QAbstractItemModel* model,
                    const QModelIndex& modelIndex)
                    const override;
  QWidget *createEditor(QWidget* parent,
                    const QStylestyleOptionViewItem& styleOption,
                    const QModelIndex & modelIndex)
                    const override;
  void setEditorData(QWidget* editor,
                    const QModelIndex& modelIndex)
                    const override;
  void updateEditorGeometry(QWidget* editor,
                    const QStylestyleOptionViewItem& styleOption,
                    const QModelIndex& modelIndex)
                    const override;
};
```

You have to override the virtual methods and add respective logic as per your project needs. You can learn more about the custom delegates and examples at the following link:

`https://doc.qt.io/qt-6/model-View-programming.html`

In this section, we learned how to create a GUI application that uses the M/V pattern. In the next section, we will discuss how it is implemented in QML.

Understanding Models and Views in QML

Just like Qt Widgets, Qt Quick also implements Models, Views, and delegates to display data. The implementation modularizes the visualization of data to empower developers to manage that data. You can change one View with another with minimal changes to the data.

To visualize data, bind the View's `model` property to a Model and the `delegate` property to a component or another compatible type.

Let's discuss the QML types available for implementing the M/V pattern in a Qt Quick application.

Views in Qt Quick

Views are containers that display data and are used for collections of items. These containers are feature-rich and can be customized to meet specific style or behavior requirements.

There is a set of standard Views provided in the basic set of Qt Quick graphical types:

- `ListView`: Lays out items in a horizontal or vertical list
- `GridView`: Lays out items in a grid manner
- `TableView`: Lays out items in a tabular form
- `PathView`: Lays out items on a path

`ListView`, `GridView`, and `TableView` inherit from the `Flickable` QML type. `PathView` inherits `Item`. The `TreeView` QML type is obsolete. Let's have a look at the inheritance of these QML types:

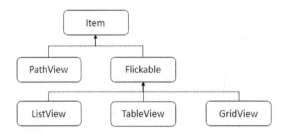

Figure 7.12 – The hierarchy of View classes in Qt Quick

The properties and behaviors are different for each QML type. They are used based on the GUI requirement. If you would like to know more about QML types, you can refer to their respective documentation. Let's explore Models in Qt Quick in the next section.

Models in Qt Quick

Qt provides several convenience QML types to implement the M/V pattern. The modules provide very simple Models without requiring the creation of custom Model classes in C++. Examples of such convenience classes are `ListModel`, `TableModel`, and `XmlListModel`.

The `QtQml.Models` module provides the following QML types for defining data Models:

- `ListModel` defines a free-form list data source.
- `ListElement` defines a data item in a `ListModel`.
- `DelegateModel` encapsulates a Model and delegate.
- `DelegateModelGroup` encapsulates a filtered set of visual data items.
- `ItemSelectionModel` inherits `QItemSelectionModel` and it keeps track of a View's selected items.
- `ObjectModel` defines a set of items to be used as a Model.
- `Instantiator` dynamically instantiates objects.
- `Package` describes a collection of named items.

To use the aforementioned QML types in your Qt Quick application, import the module with the following line:

```
import QtQml.Models
```

Let's discuss the readymade Models available in Qt Quick. `ListModel` is a simple container of `ListElement` definitions that contain data roles. It is used with `ListView`. `Qt.labs.qmlmodels` provides experimental QML types for models. These Models can be used for quick prototyping and displaying very simple data. The `TableModel` type stores JavaScript/JSON objects as data for a table Model and uses it with `TableView`. You can use these experimental types by importing the module as follows:

```
import Qt.labs.qmlmodels
```

If you want to create a Model from XML data, then you can use `XmlListModel`. It can be used as a Model with Views such as `ListView`, `PathView`, and `GridView`. To use this Model, you have to import the module as follows:

```
import QtQuick.XmlListModel
```

You can use `ListModel` and `XmlListModel` with `TableView` to create one column in `TableView`. To handle multiple rows and columns, you can use `TableModel` or you can create a custom C++ Model by subclassing `QAbstractItemModel`.

You can also use `Repeater` with Models. An integer can be used as a Model that defines the number of items. In that case, the Model does not have any data roles. Let's create a simple example that uses `ListView` and a `Text` item as delegate components:

```qml
import QtQuick
import QtQuick.Window
Window {
    width: 640
    height: 480
    visible: true
    title: qsTr("Simple M/V Demo")
    ListView {
        anchors.fill: parent
        model: 10
        delegate: itemDelegate
    }
    Component {
        id: itemDelegate
        Text { text: "  Item :  " + index }
    }
}
```

In the preceding example, we have used an **integer-based model**. We created a simple delegate, which is a text element. For a simpler explanation, we have not used a complex delegate component. You can also directly use `Text` as a delegate without using a component.

Now, let's explore how to use `ListModel` with `ListView`. `ListModel` is a simple hierarchy of types specified in QML. The available roles are specified by the `ListElement` properties. Let's create a simple application using `ListModel` with `ListView`.

Let's say you want to create a simple address book application. You may need a few fields for a contact. In the following code snippet, we have used a `ListModel` that contains the names, phone numbers, and email addresses of some contacts:

```
ListModel {
    id: contactListModel
    ListElement {
        name: "John" ; phone: "+1 1234567890" ;
        email: "john@abc.com"
    }
    ListElement {
        name: "Michael" ; phone: "+44 213243546" ;
        email: "michael@abc.com"
    }
    ListElement {
        name: "Robert" ; phone: "+61 5678912345" ;
        email: "robert@xyz.com"
    }
    ListElement {
        name: "Kayla" ; phone: "+91 9876554321" ;
        email: "kayla@xyz.com"
    }
}
```

We have now created the Model. Next, we have to display it using a delegate. So, let's modify the delegate component we created earlier with three `Text` elements. Based on your need you can create complex delegate types with icons, texts, or custom types. You can add a highlighted item and update the background based on focus. You need to provide a delegate to a View to visually represent an item in a list:

```qml
Component {
    id: contactDelegate
    Row {
        id: contact
        spacing: 20
        Text { text: " Name: " + name; }
        Text { text: " Phone no: " + phone }
        Text { text: " Email ID: " + email }
    }
}
ListView {
    anchors.fill: parent
    model: contactListModel
    delegate: contactDelegate
}
```

In the preceding example, we used `ListElement` with `ListModel`. The View displays each item as per the template defined by the delegate. Items in a Model can be accessed through the `index` property or through the item's properties.

You can learn more about different types of Models and how to manipulate Model data at the following link:

`https://doc.qt.io/qt-6/qtquick-modelviewsdata-modelview.html`

In this section, you learned about M/V in QML. You can experiment with the custom Models and delegates and create a personalized View. Have a look at your phone book or recent call list on your cellphone and try to implement it. In the next section, you will learn how to integrate the QML frontend with a C++ Model.

Using C++ Models with QML

So far, we have discussed how to use Models and Views in Qt Widgets and QML. But in most modern applications, you will require Models written in C++ and a frontend written in QML. Qt allows us to define Models in C++ and then access them inside QML. This is convenient for exposing existing C++ data Models or otherwise complex datasets to QML. Native C++ is always the right choice for complex logical operations. It can outperform logic written in QML with JavaScript.

There are many reasons why you should create a C++ Model. C++ is type-safe and compiled into object code. It increases the stability of your application and reduces the number of bugs. It is flexible and can offer more features than the QML types. You can integrate with your existing code or with a third-party library that is written in C++.

You can define a C++ Model using the following classes:

- `QStringList`
- `QVariantList`
- `QObjectList`
- `QAbstractItemModel`

The first three classes are beneficial for exposing simpler datasets. `QAbstractItemModel` offers a more flexible solution to create complex Models. `QStringList` contains a list of `QString` instances and provides the contents of the list via the `modelData` role. Similarly, `QVariantList` contains a list of `QVariant` types and provides the contents of the list via the `modelData` role. If a `QVariantList` changes, then you must reset the Model. `QObjectList` embeds a list of `QObject*` that provides the properties of the objects in the list as roles. The `QObject*` is accessible as the `modelData` property. For convenience, the properties of the object can be accessed directly in the delegate's context.

Qt also provides C++ classes to handle SQL data Model such as `QSqlQueryModel`, `QSqlTableModel`, and `QSqlRelationalTableModel`. `QSqlQueryModel` offers a read-only Model based on an SQL query. These classes reduce the need to run SQL queries for basic SQL operations such as insert, create, or update. These classes are derived from `QAbstractTableModel` and make it easy to present data from a database in a View class.

You can learn more about different types of C++ Models by visiting the following link:

`https://doc.qt.io/qt-6/qtquick-modelviewsdata-cppmodels.html`

In this section, we discussed C++ Models and why to use them. Now, you can fetch data from a C++ backend and present it in a UI developed in QML. In the next section, we will create a simple Qt Quick application using the aforementioned concept and explain how to use them inside QML.

Creating a simple M/V application with Qt Quick

In earlier sections, we discussed Qt's Model-View-Delegate framework. You learned how to create a custom Model and delegate and how to use a C++ Model. But you must be wondering how to integrate with our QML frontend. In this section, we will create a C++ Model and expose it to the QML engine. We will also discuss how to register a custom Model as a QML type.

Let's create an application that fetches a Model from the C++ code and displays it in a Qt Quick-based application:

```cpp
#include <QGuiApplication>
#include <QQmlApplicationEngine>
#include <QQmlContext>
#include <QStringListModel>
int main(int argc, char *argv[])
{
    QGuiApplication app(argc, argv);
    QQmlApplicationEngine engine;
    QStringList stringList;
    stringList << "Item 1" << "Item 2" << "Item 3"
            <<"Item 4";
    engine.rootContext()->setContextProperty("myModel",
        QVariant::fromValue(stringList));
    const QUrl url(QStringLiteral("qrc:/main.qml"));
    engine.load(url);
    return app.exec();
}
```

In the preceding code snippet, we have created a simple Model based on `QStringList`. The string list contains four different strings. We have exposed the Model to the QML engine using `setContextProperty()`. Now, let's use the Model inside our QML file:

```qml
import QtQuick
import QtQuick.Window
Window {
```

```
    width: 640
    height: 480
    visible: true
    title: qsTr("QML CPP M/V Demo")
    ListView {
        id: listview
        width: 120
        height: 200
        model: myModel
        delegate: Text { text: modelData }
    }
}
```

The preceding example uses `QQmlContext::setContextProperty()` to set Model values directly in a QML component. An alternative to this is to register the C++ Model class as a QML type as follows:

```
qmlRegisterType<MyModel>("MyModel",1,0,"MyModel");
```

The preceding line will allow the Model classes to be created directly as QML types within QML files. The first field is the C++ class name, then comes the desired package name, then the version number, and the last parameter is the type name in QML. You can import it into your QML file with the following line:

```
Import MyModel 1.0
```

Let's create an instance of MyModel inside our QML file as shown here:

```
MyModel {
    id: myModel
}
ListView {
    width: 120
    height: 200
    model: myModel
    delegate: Text { text: modelData }
}
```

You can also use Models with `QQuickView` using `setInitialProperties()` as shown in the following code:

```
QQuickView view;
view.setResizeMode(QQuickView::SizeRootObjectToView);
view.setInitialProperties({
                {"myModel",QVariant::fromValue(myModel)}});
view.setSource(QUrl("qrc:/main.qml"));
view.show();
```

In the preceding code snippet, we used `QQuickView` to create a UI and passed a custom C++ Model to the QML environment.

In this section, we learned how to integrate a simple C++ Model with QML. You can add signals and properties to extend the functionalities of your custom classes. Next, let's summarize our learnings in this chapter.

Summary

In this chapter, we took a look at the core concepts of the Model-View-Delegate pattern in Qt. We explained how it is different from the traditional MVC pattern. We discussed different ways of using M/V and the convenience classes available in Qt. We learned how to apply the M/V concept in Qt Widgets as well as in Qt Quick. We discussed how to integrate a C++ Model with QML Views. We also created a few examples and implemented the concepts in our Qt application. You can now create your own Model, delegate, and Views. I hope you have understood the importance of the framework and the solid reasons for using it to meet your requirements.

In *Chapter 8, Graphics and Animations*, we will learn about the graphics framework and how to add animations to your Qt Quick project.

8
Graphics and Animations

In this chapter, you will learn the fundamentals of Qt's graphics framework and how to render graphics on a screen. You will understand how general drawing is done in Qt. We will begin by discussing 2D graphics using **QPainter**. We will explore how to draw different shapes using a painter. Then you will learn about the Graphics View architecture used by **QGraphicsView** and **QGraphicsScene**. Later, we will discuss the **Scene Graph** mechanism used by Qt Quick. In this chapter, you will also learn how to make the user interface more interesting by adding animations and states.

In this chapter, we will discuss the following:

- Understanding Qt's graphics framework
- `QPainter` and 2D graphics
- The Graphics View framework
- OpenGL implementation
- Qt Quick scene graph
- Animation in QML
- State machines in Qt

By the end of this chapter, you will understand the graphics framework used by Qt. You will be able to draw onscreen and add animations to your UI elements.

Technical requirements

The technical requirements for this chapter include minimum versions of Qt 6.0.0 and Qt Creator 4.14.0 installed on the latest version of a desktop platform such as Windows 10, Ubuntu 20.04, or macOS 10.14.

All the code used in this chapter can be downloaded from the following GitHub link: `https://github.com/PacktPublishing/Cross-Platform-Development-with-Qt-6-and-Modern-Cpp/tree/master/Chapter08`.

> **Important note**
>
> The screenshots used in this chapter are taken from the Windows platform. You will see similar screens based on the underlying platforms on your machine.

Understanding Qt's graphics framework

Qt is one of the most popular frameworks for GUI applications. Developers can build awesome cross-platform GUI applications using Qt without worrying about the underlying graphics implementation. The Qt **Rendering Hardware Interface** (**RHI**) interprets graphics instructions from Qt applications to the available graphics APIs on the target platform.

RHI is the abstract interface for hardware-accelerated graphics APIs. The most important class in the `rhi` module is `QRhi`. The `QRhi` instance is supported by a backend for the specific graphics API. The selection of the backend occurs at runtime and is decided by the application or library that creates the `QRhi` instance. You can add the module by adding the following line into your project file:

```
QT += rhi
```

Different types of graphics APIs supported by RHI are as follows:

- **OpenGL**
- **OpenGL ES**
- **Vulkan**
- **Direct3D**
- **Metal**

Figure 8.1 shows the major layers of the graphics stack in the Qt graphics framework:

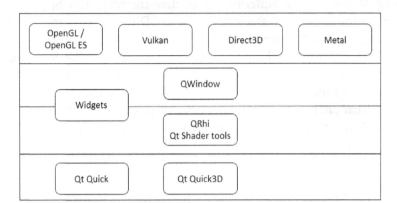

Figure 8.1 – Major layers of the Qt 6 graphics stack

Let's get familiar with the graphics APIs shown in the preceding diagram. **OpenGL** is the most popular graphics API with cross-language and cross-platform application support. It is used to interact with the GPU, to achieve hardware-accelerated rendering. **OpenGL ES** is a flavor of the OpenGL API intended for embedded devices. It allows the rendering of advanced 2D and 3D graphics on embedded and mobile devices. **OpenGL ES on iOS devices** is also known as **EAGL**. OpenGL ES is also available on web platforms as WebGL. OpenGL and OpenGL ES are developed and maintained by the Khronos Group, a consortium of technology hardware and software companies. You can learn more about OpenGL at the following link:

https://www.opengl.org/about/

Vulkan is a new-generation graphics API that helps in creating cross-platform and high-performance applications for modern GPUs. It was created by the Khronos Group. Vulkan's explicit API design allows efficient implementations on a wide range of desktop, embedded, and mobile platforms. Qt 6 provides support for the Vulkan API. To use Vulkan, Qt applications require the LunarG Vulkan SDK. Explore more about Vulkan at the following link:

https://www.lunarg.com/vulkan-sdk/

Direct3D is a Microsoft proprietary graphics API that provides functions to render 2D and 3D graphics by utilizing underlying GPU capabilities. Microsoft Corporation created it for use on the Windows platform. It is a low-level API that can be used to draw primitives with the rendering pipeline or to perform parallel operations with the compute shader.

Direct3D exposes the advanced graphics capabilities of 3D graphics hardware, including stencil buffering, W-buffering, Z-buffering, perspective-correct texture mapping, spatial anti-aliasing, programmable HLSL shaders, and effects. Direct3D's integration with other DirectX technologies allows it to provide several features comprising video mapping, hardware 3D rendering in 2D overlay planes, and even sprites and allowing the use of 2D and 3D graphics in interactive media ties. Direct3D is intended to virtualize 3D hardware interfaces in general. In contrast, OpenGL is intended to be a 3D hardware-accelerated rendering system that can be emulated in software. These two APIs are fundamentally designed in two distinct ways. The following link provides further insight into Direct3D:

```
https://docs.microsoft.com/en-in/windows/win32/getting-
started-with-direct3d
```

Metal is Apple's low-level computer graphics API, which offers near-direct access to the **graphics processing unit (GPU)**, allowing you to optimize the graphics and compute capacity of your iOS, macOS, and tvOS apps. It also has a low-overhead architecture that includes pre-compiled GPU shaders, fine-grained resource management, and multithreading support. Before the announcement of Metal, Apple provided OpenGL for macOS and OpenGL ES for iOS, but there was a performance issue due to the highly abstracted hardware. Metal, on the other hand, has better performance than OpenGL thanks to its Apple-specific API. Metal enables a whole new generation of professional graphics output by supporting up to 100 times more draw calls than OpenGL. You can read more about Metal at the following link:

```
https://developer.apple.com/documentation/metal
```

In this section, we got familiar with Qt's graphics framework and RHI. You now have a basic understanding of this framework. In the next section, we will go further and discuss 2D graphics using QPainter.

QPainter and 2D graphics

Qt comes with an advanced windowing, painting, and typography system. The most important classes in the Qt GUI module are `QWindow` and `QGuiApplication`. This module includes classes for 2D graphics, imaging, fonts, and advanced typography. Additionally, the GUI module comes with classes for integrating windowing systems, OpenGL integration, event handling, 2D graphics, basic imaging, fonts, and text. Qt's user interface technologies use these classes internally, but they can directly be used to write applications that use low-level OpenGL graphics APIs.

Depending on the platform, the `QWindow` class supports rendering with OpenGL and OpenGL ES. Qt includes the `QOpenGLPaintDevice` class, which allows the use of OpenGL accelerated `QPainter` rendering and several convenience classes. These convenience classes simplify writing code in OpenGL by hiding the complexities of extension handling and the differences between OpenGL ES 2.0 and desktop OpenGL. `QOpenGLFunctions` is a convenience class that provides cross-platform access to the OpenGL ES 2.0 functions on desktop OpenGL without the need to manually resolve the OpenGL function pointers.

To make use of these APIs and classes on a qmake-based application, you have to include the `gui` module in your project file (`.pro`) as follows:

```
QT += gui
```

If you are using a *Cmake*-based build system, then add the following to the `CMakeLists.txt` file:

```
find_package(Qt6 COMPONENTS Gui REQUIRED)
target_link_libraries(mytarget PRIVATE Qt6::Gui)
```

The `QPainter` class, primarily used for drawing operations, provides an API for various tasks such as drawing vector graphics, text, and images onto different surfaces, or `QPaintDevice` instances, including `QImage`, `QOpenGLPaintDevice`, `QWidget`, and `QPrinter`. For Qt Widgets user interfaces, Qt uses a software renderer.

The following are Qt GUI's high-level drawing APIs:

- Paint system
- Coordinate system
- Drawing and filling

We will explore these APIs in the following sections.

Understanding the paint system

Qt's paint system provides several convenience classes for drawing on the screen. The most important classes used are QPainter, QPaintDevice, and QPaintEngine. You can use QPainter to paint on widgets and other paint devices. This class can be used to draw things from simple lines to complex shapes such as **pies** and **chords**. It is also used to draw **pixmaps** and **texts**. If the paint device is a widget, then use QPainter inside the paintEvent() function or inside a function invoked by a function called by paintEvent(). QPaintDevice is the base class of the objects that allow 2D drawing by using a QPainter instance. QPaintEngine provides the interface that defines how QPainter paints to a specified device on a specified platform. The QPaintEngine class is an abstract class that is used internally by QPainter and QPaintDevice.

Let's have a look at the hierarchy of painting-related classes to get a better idea of how to choose the right classes while using the paint system.

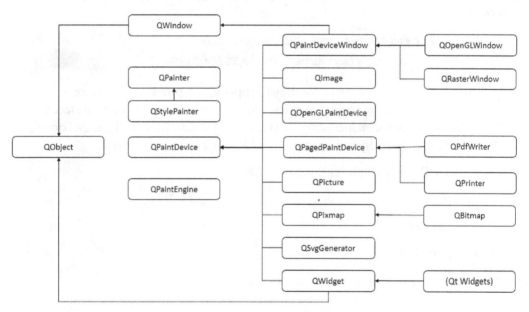

Figure 8.2 – The hierarchy of paint classes in Qt

The preceding hierarchical approach illustrates that all drawing approaches follow the same mechanism. So, it is easy to add provisions for new features and provide default implementations for unsupported ones.

Let's discuss the coordinate system in the next section.

Using the coordinate system

The `QPainter` class controls the coordinate system. It forms the basis of Qt's painting system along with the `QPaintDevice` and `QPaintEngine` classes. The default coordinate system of a paint device has its origin in the top-left corner. The primary function of `QPainter` is to perform drawing operations. While the `QPaintDevice` class is an abstraction of a two-dimensional space, which can be painted on using `QPainter`, the `QPaintEngine` class offers a painter with the interface to draw on different types of devices. The `QPaintDevice` class is the base class of objects that can be painted, which inherits its drawing capabilities from the `QWidget`, `QImage`, `QPixmap`, `QPicture`, and `QOpenGLPaintDevice` classes.

You can learn more about the coordinate system in the following documentation:

`https://doc.qt.io/qt-6/coordsys.html`

Drawing and filling

`QPainter` provides a painter with highly optimized functions for most of the drawing requirements on the GUI. It can draw various types of shapes ranging from simple graphical primitives (such as `QPoint`, `QLine`, `QRect`, `QRegion`, and `QPolygon` classes) to complex shapes such as vector paths. The vector paths are represented by the `QPainterPath` class. `QPainterPath` works as a container for painting operations, allowing graphical shapes to be constructed and reused. It can be used for filling, outlining, and clipping. `QPainter` can also draw aligned text and pixmaps. To fill the shapes drawn by `QPainter`, you can use the `QBrush` class. It has color, style, texture, and gradient attributes and is defined with color and style.

In the next section, we will use the APIs discussed so far to draw using `QPainter`.

Drawing with QPainter

`QPainter` has several convenience functions to draw most primitive shapes, such as `drawLine()`, `drawRect()`, `drawEllipse()`, `drawArc()`, `drawPie()`, and `drawPolygon()`. You can fill the shapes using the `fillRect()` function. The `QBrush` class describes the fill pattern of shapes drawn by `QPainter`. A brush can be used to define the style, color, gradient, and texture.

Let's look at the following `paintEvent()` function where we have used `QPainter` to draw text and different shapes:

```cpp
void PaintWindow::paintEvent(QPaintEvent *event)
{
    QPainter painter;
    painter.begin(this);
    //draws a line
    painter.drawLine(QPoint(50, 50), QPoint(200, 50));
    //draws a text
    painter.drawText(QPoint(50, 100), "Text");
    //draws an ellipse
    painter.drawEllipse(QPoint(100,150),50,20);
    //draws an arc
    QRectF drawingRect(50, 200, 100, 50);
    int startAngle = 90 * 16;
    int spanAngle = 180 * 16;
    painter.drawArc(drawingRect, startAngle, spanAngle);
    //draws a pie
    QRectF drawingRectPie(150, 200, 100, 50);
    startAngle = 60 * 16;
    spanAngle = 70 * 16;
    painter.drawPie(drawingRectPie, startAngle, spanAngle);

    painter.end();

    QWidget::paintEvent(event);
}
```

In the preceding example, we have created a `QPainter` instance and painted a line, text, ellipse, arc, and pie using the available default drawing functions. When you add the preceding code into your custom class and run the project, you will see the following output:

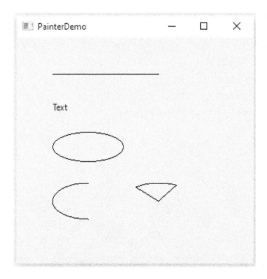

Figure 8.3 – Output of drawing with QPainter example

Qt provides several offscreen drawing classes, each with its own set of advantages and disadvantages. QImage, QBitmap, QPixmap, and QPicture are the classes involved. In most cases, you must choose between QImage and QPixmap.

The QImage class in Qt allows for easy image reading, writing, and manipulation. QImage is the class to use if you're working with resources, combining multiple images, and doing some drawing:

```
QImage image(128, 128, QImage::Format_ARGB32);
QPainter painter(&image);
```

The first line creates an image that's 128 pixels square, encoding each pixel a 32-bit integer – 8 bits for each channel of opacity, red, green, and blue. The second line creates a QPainter instance that can draw on the QImage instance. Next, we perform the drawing you just saw in the previous section, and when we're done, we write the image to a PNG file, with the following line:

```
image.save("image.png");
```

QImage supports several image formats, including PNG and JPEG. QImage also has a load method, where you can load an image from a file or resource.

The QBitmap class is a monochromatic offscreen paint device that provides a pixmap with a depth of 1 bit. The QPixmap class provides an offscreen paint device. The QPicture class is a paint device that serializes QPainter commands.

You can also use the QImageReader and QImageWriter classes to have more fine-grained control over how images are loaded and saved. To add support for image formats other than those provided by Qt, image format plugins can be created using QImageIOHandler and QImageIOPlugin. The QPainterPath class helps in drawing different graphical shapes that can be created and reused. The following code snippet demonstrates how to use QPainterPath:

```cpp
void MyWidget:: paintEvent(QPaintEvent *event)
{
    QPainter painter(this);
    QPolygon polygon;
    polygon << QPoint(100, 185) << QPoint(175, 175)
            << QPoint(200, 110) << QPoint(225, 175)
            << QPoint(300, 185) << QPoint(250, 225)
            << QPoint(260, 290) << QPoint(200, 250)
            << QPoint(140, 290) << QPoint(150, 225)
            << QPoint(100, 185);
    QBrush brush;
    brush.setColor(Qt::yellow);
    brush.setStyle(Qt::SolidPattern);
    QPen pen(Qt::black, 3, Qt::DashDotDotLine,
            Qt::RoundCap, Qt::RoundJoin);
    painter.setPen(pen);

    QPainterPath path;
    path.addPolygon(polygon);
    painter.drawPolygon(polygon);
    painter.fillPath(path, brush);
    QWidget:: paintEvent(event);
}
```

In the preceding code, we have created a custom-drawn polygonal object with the desired painter path.

> **Note**
>
> Please note that while doing a paint operation, ensure that there is no delay between painting the background and painting the content. Otherwise, you will notice flickering on the screen if the delay is more than 16 milliseconds. You can avoid this by rendering the background into a pixmap, then painting the content onto that pixmap. Finally, you can draw that pixmap onto the widget. This approach is known as **double buffering**.

In this section, we have learned not only how to draw an image on the screen, but also how to draw it off the screen and save it as an image file. In the next section, we will learn about the basics of the Graphics View framework.

Introducing the Graphics View framework

The Graphics View framework is a powerful graphics engine that allows you to visualize and interact with a large number of custom-made 2D graphical items. If you are an experienced programmer, you can use the graphics view framework to draw your GUI and have it animated completely manually. To draw hundreds or thousands of relatively lightweight customized items at once, Qt provides a separate view framework, the Graphics View framework. You can make use of the Graphics View framework if you are creating your own widget set from scratch, or if you have a large number of items to display on the screen at once, each with its own position and data. This is especially important for applications that process and display a large amount of data, such as geographic information systems or computer-aided design software.

Graphics View offers a surface for managing as well as interacting with a multitude of custom-created 2D graphical items, and a view widget for visualizing the items, with zooming and rotation support. The framework consists of an event propagation architecture that enables interaction capabilities for the scene's items. These items respond to key events; mouse press, move, release, and double-click events; as well as tracking mouse movement. Graphics View employs a **Binary Space Partitioning** (BSP) tree to provide very fast item discovery, allowing it to visualize large scenes in real time, even when there are millions of items.

The framework follows an item-based `approach` to model/view programming. It comprises three components, **scene**, **view**, and **item**. Multiple views can use the same scene and the scene can contain multiple items. The convenience classes provided by Qt to implement the Graphics View framework are `QGraphicsScene`, `QGraphicsView`, and `QGraphicsItem`.

QGraphicsItem exposes an interface that your subclass can override to manage mouse and keyboard events, drag and drop, interface hierarchies, and collision detection. Each item has its own local coordinate system, and helper functions allow you to quickly transform an item's coordinates to the scene's coordinates. The Graphics View framework displays the contents of a QGraphicsScene class using one or more QGraphicsView instances. To see different parts of the scene, you can attach multiple views to the same scene, each with its own translation and rotation. Because the QGraphicsView widget is a scroll area, you can also attach scroll bars to the view and allow the user to scroll around it. The view receives keyboard and mouse input, generates scene events for the scene, and dispatches those scene events to the scene, which then dispatches those same events to the scene's items. Previously, the framework was preferred for games development.

> **Important note**
>
> We will skip the details about the usages of the framework and examples as it lost its popularity after Qt Quick 2 came into existence. Qt Quick 2 comes with the Scene Graph API, which provides most of the functionalities that were earlier offered by the Graphics View framework. If you'd still like to learn more about the Graphics View framework, you can read the following documentation:
>
> https://doc.qt.io/qt-6/graphicsview.html

In this section, we discussed Qt's Graphics View framework. In the next section, we will learn about OpenGL integration with Qt.

Understanding the Qt OpenGL module

Qt Quick and Qt Widgets are the two main approaches to **user interface** (**UI**) development in Qt. They exist to support various types of UIs and are built on separate graphics engines that have been optimized for each of these. It is possible to combine OpenGL graphics API code with both of these UI types in Qt. This is useful when the application contains its own OpenGL-dependent code or when integrating with a third-party OpenGL-based renderer. The OpenGL/OpenGL ES XML API Registry is used to generate the OpenGL header.

The Qt OpenGL module is intended for use with applications that require OpenGL access. The convenience classes in the Qt OpenGL module help developers build applications more easily and faster. This module is responsible for maintaining compatibility with Qt 5 applications and Qt GUI. QOpenGLWidget is a widget that can add OpenGL scenes to UIs that use QWidget.

With the introduction of Qt RHI as the rendering foundation in Qt, most classes denoted by QOpenGL have been moved to the Qt OpenGL module in Qt 6. The classes are still usable and fully supported for applications that rely solely on OpenGL. They are no longer considered essential because Qt has been extended to support other graphics APIs, such as Direct3D, Metal, and Vulkan, in its foundation.

Existing application code will mostly continue to work, but it should now include Qt OpenGL in project files, as well as the headers if they were previously included indirectly via Qt GUI.

Qt 6 no longer directly employs OpenGL-compatible GLSL source snippets. Shaders are instead written in Vulkan-style GLSL, reflected and translated to other shading languages, and packaged into a serializable QShader object that QRhi can consume.

The shader preparation pipeline in Qt 6 is the following:

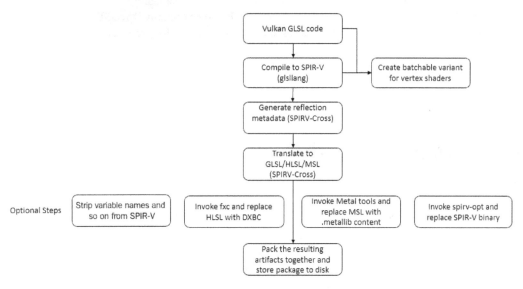

Figure 8.4 – Illustration of the shader preparation pipeline as described in the Qt blog

With Qt 6.1, Qt Data Visualization supports only the OpenGL RHI backend. It requires the setting of the environment variable QSG_RHI_BACKEND to opengl. You can do this at the system level, or define it in main() as follows:

```
qputenv("QSG_RHI_BACKEND", "opengl");
```

Let's discuss how the framework is used with Qt Widgets in the next section.

Qt OpenGL and Qt Widgets

Qt Widgets is typically rendered by a highly optimized and accurate software rasterizer, with the final content being displayed on the screen using a method appropriate for the platform on which the application is running. However, Qt Widgets and OpenGL can be combined. The QOpenGLWidget class is the primary entry point for this. This class can be used to enable OpenGL rendering for a specific part of the widget tree, and the Qt OpenGL module's classes can be used to help with any application-side OpenGL code.

> **Important note**
>
> **ANGLE**, a third-party OpenGL ES to Direct3D translator, is no longer included with Qt 6 on Windows. For QWindow or QWidget based applications with OpenGL implementations, there are no other options but to directly call the OpenGL APIs at runtime. For Qt Quick and Qt Quick 3D applications, Qt 6 introduced support for Direct3D 11, Vulkan, and Metal, in addition to OpenGL. On Windows, the default choice remains Direct3D, therefore the removal of ANGLE is eased by having support for graphics APIs other than OpenGL as well.

In this section, we learned how to use Qt's Open GL module. Let's move on to the next section, where we'll discuss graphics in Qt Quick in detail.

Graphics in Qt Quick

Qt Quick is designed to take advantage of hardware-accelerated rendering. It will be built by default on the low-level graphics API that is most appropriate for the target platform. On Windows, for example, it will default to Direct3D, whereas on macOS, it will default to Metal. For rendering, Qt Quick applications make use of a scene graph. The scene graph renderer can make more efficient graphics calls, which improves performance. The scene graph has an accessible API that allows you to create complex but fast graphics. The Qt Quick 2D Renderer can also be used to render Qt Quick. This raster paint engine allows Qt Quick applications to be rendered on platforms that do not support OpenGL.

Qt uses the most appropriate graphics API on the target platform by default. However, it is possible to configure Qt's rendering path to use a specific API. In many cases, selecting a specific API improves performance and allows developers to deploy on platforms that support a specific graphics API. To change the render path in QQuickWindow, you can use the QRhi interface.

In the following sections, we will have a look at some functionalities that will further enhance your graphics-related skills in Qt Quick. Let's begin by discussing how we can use OpenGL in Qt Quick.

Qt OpenGL and Qt Quick

On platforms that support OpenGL, it is possible to manually select it as the active graphics API. In order to use this functionality when working with Qt Quick, the application should manually set the rendering backend to OpenGL in addition to adjusting project files and including headers.

With Qt 6, there is no direct way of OpenGL rendering using Qt Quick. The QRhi-based rendering path of the Qt Quick scene graph is now the new default. Aside from the defaults, the methods for configuring which QRhi backend and thus which graphics API to use remain largely unchanged from Qt 5.15. One key difference in Qt 6 is improved API naming. Now, you can set the RHI backend by calling the `QQuickWindow::setGraphicsApi()` function, whereas earlier this was achieved by calling the `QQuickWindow::setSceneGraphBackend()` function.

You can learn more about the changes in the following article:

`https://www.qt.io/blog/graphics-in-qt-6.0-qrhi-qt-quick-qt-quick-3d`

Custom Qt Quick items using QPainter

You can also make use of `QPainter` in your Qt Quick application. This can be done by subclassing `QQuickPaintedItem`. With the help of this subclass, you can render content using a `QPainter` instance. To render its content, the `QQuickPaintedItem` subclass uses an indirect 2D surface by either using software rasterization or using an **OpenGL Framebuffer Object** (**FBO**). Rendering is a two-step operation. The paint surface is rasterized before drawing. However, drawing using a scene graph is significantly faster than this rasterization approach.

Let's explore the scene graph mechanism used by Qt Quick.

Understanding the Qt Quick scene graph

Qt Quick 2 employs a dedicated scene graph that is traversed and rendered using a graphics API, including OpenGL, OpenGL ES, Metal, Vulkan, or Direct 3D. Using a scene graph for graphics instead of traditional imperative painting systems (`QPainter` and similar), allows the scene to be rendered to be retained between frames and the entire set of primitives to render to be known before rendering begins. This allows for a variety of optimizations, including batch rendering to reduce state changes and discarding obscured primitives.

Let's assume a GUI comprises a list of 10 elements and each one has a different background color, text, and icon. This would give us 30 draw calls and an identical number of state changes using traditional drawing techniques. Contrarily, a scene graph reorganizes the primitives to render so that one call can draw all backgrounds, icons, and text, dropping the total number of draw calls to three. This type of batching and state change reduction can significantly improve performance on some hardware.

The scene graph is inextricably linked to Qt Quick 2 and cannot be used independently. The QQuickWindow class manages and renders the scene graph, and custom Item types can add their graphical primitives to the scene graph by calling QQuickItem::updatePaintNode().

The scene graph represents an Item scene graphically and is a self-contained structure that has enough information to render all of the items. Once configured, it can be manipulated and rendered regardless of the state of the items. On several platforms, the scene graph is even rendered on a separate render thread while the GUI thread prepares the state for the next frame.

In the following sections, we will dive deeper to improve our understanding of the scene graph structure and then learn the rendering mechanism. Further, we will be mixing the scene graph and the Native Graphics API while using Qt Quick 3D.

Qt Quick scene graph structure

The scene graph is made up of a variety of predefined node types, each of which serves a specific purpose. Although we call it a scene graph, a node tree is a more precise definition. The tree is constructed from QQuickItem types in the QML scene, and the scene is then internally processed by a renderer, which draws the scene. There is no active drawing code in the nodes themselves.

Although the node tree is mostly built internally by the existing Qt Quick QML types, users can add complete subtrees with their own content, including subtrees that represent 3D models.

- Node
- Material

QSGGeometryNode is the most important node for users. It creates customized graphics by specifying their geometry and material. The QSGGeometry class describes the shape or mesh of the graphical primitive and is used to define the geometry. It can define everything, be it a line, a rectangle, a polygon, a collection of disconnected rectangles, or a complex 3D mesh. The material defines how the pixels for a specific shape are filled. There can be multiple children for a node. The geometry nodes are rendered as per the child order and the parent nodes can be found behind their children.

The material describes how a geometry's interior in `QSGGeometryNode` is filled. It encapsulates graphics shaders for the vertex and fragment stages of the graphics pipeline and provides a great deal of flexibility in what can be done, even though the majority of Qt Quick items only use very basic materials such as solid color and texture fills.

The scene graph API is low-level and prioritizes performance over convenience. Creating the most basic custom geometries and materials from scratch requires a significant amount of code input. As a result, the API includes a few convenience classes that make the most commonly used custom nodes easily accessible.

In the next section, we will discuss how the rendering is done in a scene graph.

Rendering using a scene graph

A scene graph is internally rendered in the `QQuickWindow` class, and there is no public API to access it. However, there are a few points in the rendering pipeline where the user can insert application code. These points can be used for adding custom scene graph content or for inserting arbitrary rendering commands by calling the scene graph's graphics API (OpenGL, Vulkan, Metal, and so on) directly. The render loop determines the integration points.

There are two types of render loops in a scene graph:

- `basic` is a single-threaded renderer.
- `threaded` is a multithread renderer that renders on a different thread.

Qt tries to select an appropriate render loop based on the platform and underlying graphics capabilities. When this is not sufficient, or during testing, the environment variable `QSG_RENDER_LOOP` can be used to force the use of a specific type of renderer loop. You can find the type of render loop in use by enabling the `qt.scenegraph.general` logging category.

In most applications that use a scene graph, the rendering takes place on a separate render thread. This is done to improve multi-core processor parallelism and make better use of stall times such as waiting for a blocking swap buffer call. This provides significant performance improvements, but it limits where and when interactions with the scene graph can occur.

The following diagram depicts how a frame is rendered using the threaded render loop and OpenGL. Apart from the OpenGL context specifics, the steps are the same for other graphics APIs as well:

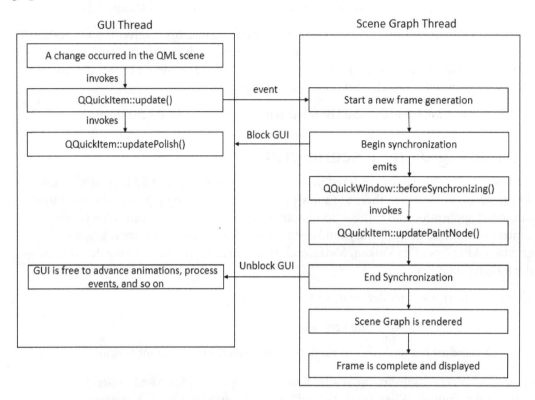

Figure 8.5 – Rendering sequence followed in a threaded render loop

Currently, the threaded renderer is used by default on Windows with Direct3D 11 or higher. You can force the use of the threaded renderer by setting QSG_RENDER_LOOP to threaded in the environment. However, the threaded render loop depends on the graphics API implementation for throttling. When building with Xcode 10 or later on macOS and OpenGL, the threaded render loop is not supported. For Metal, there are no such limitations.

If your system is not capable of providing Vsync-based throttling, then use the basic render loop by setting the environment variable QSG_RENDER_LOOP to basic. The following steps describe how a frame is rendered in a basic or non-threaded render loop:

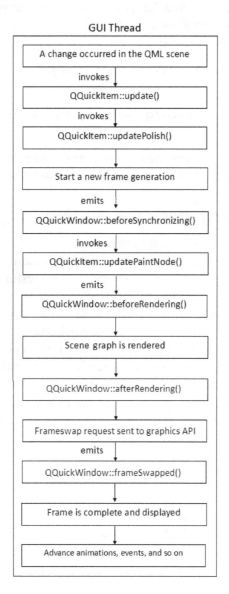

Figure 8.6 – Rendering sequence followed in a non-threaded render loop

When the platform's standard OpenGL library is not used, then by default the non-threaded render loop is used on OpenGL-enabled platforms. This is primarily a preventive strategy for the latter because not all the combinations of OpenGL drivers and windowing systems have been verified. You may consider writing your code as if you are using the threaded renderer even if you are using the non-threaded render loop because otherwise, your code won't be portable.

To find further information on the workings of the scene graph renderer, you may visit the following link:

```
https://doc-snapshots.qt.io/qt6-dev/qtquick-visualcanvas-
scenegraph.html
```

In this section, you got to know about the rendering mechanism behind the scene graph. In the next section, we will discuss how to mix a scene graph with the Native Graphics API.

Using a scene graph with the Native Graphics

The scene graph provides two methods for mixing the scene graph with the Native Graphics APIs. The first approach is by directly issuing commands to the underlying graphics engine, and the second approach is by generating a textured node in the scene graph. Applications can make OpenGL calls directly into the same context as the scene graph by connecting to the `QQuickWindow::beforeRendering()` and `QQuickWindow::afterRendering()` signals. Applications using APIs such as Metal or Vulkan can request native objects, such as the scene graph's command buffer, through `QSGRendererInterface`. Then the user can render content either within or outside of the Qt Quick scene. The advantage of mixing the two is that no additional framebuffer or memory is required to execute the rendering, and a potentially costly texturing step is avoided. The disadvantage is that Qt Quick chooses when to invoke the signals. The OpenGL engine is only allowed to draw during that time.

Beginning with Qt 6.0, direct use of the Native Graphics API must be invoked before the calls to the `QQuickWindow::beginExternalCommands()` and `QQuickWindow::endExternalCommands()` functions. This approach is identical to `QPainter::beginNativePainting()`, and it serves the same purpose. It allows the scene graph to identify any cached state or assumptions about the state inside the presently recorded render pass. If anything exists, then it becomes invalid as the code may have changed it by interacting directly with the Native Graphics API.

> **Important note**
>
> When combining OpenGL content with scene graph rendering, it is crucial that the application doesn't leave the OpenGL context with buffers bound, attributes enabled, or specific values in the stencil buffer, or something similar. If you forget this, then you will see unexpected behavior. The custom rendering code must be thread-aware.

The scene graph also provides support with several logging categories. These are useful in finding the root cause of performance issues and bugs. The scene graph features an adaptation layer in addition to the public API. The layer allows you to implement certain hardware-specific adaptations. It has an internal and proprietary plugin API that allows hardware adaption teams to get the most out of their hardware.

> **Important note**
>
> If you are observing graphics-related issues or to find which type of rendering loop or graphics API is currently used, start the application by setting the environment variable `QSG_INFO` to 1 or by enabling at least `qt.scenegraph.general` and `qt.rhi.*`. During initialization, this will print some crucial information required to debug the graphics issues.

3D graphics with Qt Quick 3D

Qt Quick 3D is a Qt Quick add-on that provides a high-level API for creating 3D content and 3D user interfaces. It extends the Qt Quick scene graph, allowing you to integrate 3D content into 2D Qt Quick applications. Qt Quick 3D is a high-level API for creating 3D content and 3D user interfaces on the Qt Quick platform. Rather than relying on an external engine, which introduces syncing issues and additional layers of abstraction, we provide spatial content extensions to the existing Qt Quick scene graph, as well as a renderer for that extended scene graph. It is also possible to mix Qt Quick 2D and 3D content when using the spatial scene graph.

The following `import` statement in your `.qml` file can be used to import the QML types into your application:

```
import QtQuick3D
```

In addition to the base Qt Quick 3D model, additional functionality is provided by the following module imports:

```
import QtQuick3D.Effects
import QtQuick3D.Helpers
```

Qt Quick 3D is available for purchase under a commercial license. When building from source, make sure the modules and tools from the `qtdeclarative` and `qtshadertools` repositories are built first, as Qt Quick 3D cannot be used without them.

Let's discuss shader tools and shader effects in the next section.

Shader effects

For importing shaders into 3D scenes, Qt Quick 3D has its own framework. **Shader effects** enable the full, raw power of a graphics processing unit to be directly utilized via vertex and fragment shaders. Too many shader effects can result in increased power consumption and sometimes slow performance, but when used sparingly and carefully, a shader can allow complex and visually appealing effects to be applied to a visual object.

Both shaders are bound to the `vertexShader` and `fragmentShader` properties. Every shader's code requires a `main() {...}` function, which is executed by the GPU. A variable with the prefix `qt_` is provided by Qt. To understand the variables in shader code, have a look at the OpenGL API reference document.

When working with `ShaderEffect` or subclassing `QSGMaterialShader` in QML applications using Qt Quick, the application must provide a baked shader pack in the form of a `.qsb` file. The Qt Shader Tools module includes a command-line tool called **qsb**. It incorporates third-party libraries such as **glslang** and **SPIRV-Cross**, as well as external tools such as **fxc** and **spirv-opt**, and generates `.qsb` files. The `ShaderEffect` QML type and `QSGMaterial` subclasses, in particular, can make use of qsb output. It can also be used to inspect the contents of a `.qsb` package. The input file extension is used to determine the type of shader. As a result, the extension has to be one of the following:

- `.vert` – Vertex shaders

- `.frag` – Fragment shaders

- `.comp` – Compute shaders

The example assumes `myeffect.vert` and `myeffect.frag` contain Vulkan-style GLSL code, processed by the qsb tool in order to generate the `.qsb` files. Now we convert that Vulkan-Style shader with qsb via the following command:

```
>qsb --glsl 100es,120,150 --hlsl 50 --msl 12 -o <Output_File.
qsb> <Input_File.frag>
```

You can see an example of using the preceding syntax in the following command:

```
>C:\Qt\6.0.2\mingw81_64\bin>qsb --glsl 100es,120,150 --hlsl 50
--msl 12 -o myeffect.frag.qsb myeffect.frag
```

It is not necessary to specify both vertexShader and fragmentShader. Many ShaderEffect implementations will only provide a fragment shader in practice, instead of relying on the built-in vertex shader.

You can learn more about the shader tools at the following link:

https://doc.qt.io/qt-6/qtshadertools-qsb.html

Let's use shader effects in an example:

```
import QtQuick
import QtQuick.Window
Window {
    width: 512
    height: 512
    visible: true
    title: qsTr("Shader Effects Demo")
    Row {
        anchors.centerIn: parent
        width: 300
        spacing: 20
        Image {
            id: originalImage
            width: 128; height: 94
            source: "qrc:/logo.png"
        }
        ShaderEffect {
            width: 160; height: width
            property variant source: originalImage
            vertexShader: "grayeffect.vert.qsb"
            fragmentShader: "grayeffect.frag.qsb"
        }
    }
}
```

In the preceding example, we arranged two images in a row. The first one is the original image and the second one is the image with the shader effect.

In this section, you learned about different types of shader effects in Qt Quick and how to use the qsb tool to create compatible fragment files. In the next section, you will learn how to draw using `Canvas`.

Using the Canvas QML type

Canvas allows you to draw straight and curved lines, simple and complex shapes, graphs, and graphic images that have been referenced. Text, colors, shadows, gradients, and patterns can also be added, as well as low-level pixel operations. You can save a `Canvas` output as an image. It provides a 2D canvas that uses a `Context2D` object for drawing and implements a paint signal handler.

Let's have a look at the following example:

```
import QtQuick
import QtQuick.Window
Window {
    width: 512
    height: 512
    visible: true
    title: qsTr("Canvas Demo")
    Canvas {
        id: canvas
        anchors.fill: parent
        onPaint: {
            var context = getContext("2d")
            context.lineWidth = 2
            context.strokeStyle = "red"
            context.beginPath()
            context.moveTo(100,100)
            context.lineTo(250,100)
            context.lineTo(250,150)
            context.lineTo(100,150)
            context.closePath()
            context.stroke()
```

```
            }
        }
    }
```

In the preceding example, first, we got the context from getContext("2d"). Then we drew a rectangle with a red border. The output looks as follows:

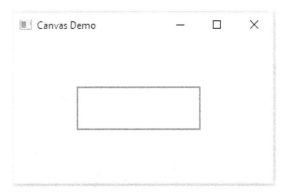

Figure 8.7 – Output of sample application using Canvas to draw a rectangle

In this section, you got familiar with drawing using Canvas. In the next section, we will discuss particle systems in Qt Quick.

Understanding particle simulations

Using particle systems, you can simulate effects such as explosions, fireworks, smoke, fog, and wind. Qt Quick includes a particle system that enables these types of complex, 2D simulations, including support for environmental effects such as gravity and turbulence. Particles are most commonly used in games to add subtle and visually appealing effects to currently selected items in lists or activity notifiers.

ParticleSystem, Painters, Emitters, and Affectors are the four main QML types in this particle system. The ParticleSystem system includes painter, emitter, and affector types. The ParticleSystem type connects all of these types and manages the shared timeline. They must all share the same ParticleSystem in order to interact. Subject to this constraint, you may have as many particle systems as you want, so the logical separation is to have one ParticleSystem type for all the types with which you want to interact, or just one if the number of types is small and easily controlled.

To use ParticleSystem, import the module with the following line:

```
import QtQuick.Particles
```

The emitter produces particles. The emitter can no longer change a particle after it has been emitted. You can use `affectors` type to influence particles after they have been emitted.

Each type of `affector` affects particles differently:

- `Age`: Modifies the particle's lifespan
- `Attractor`: Draws particles towards a certain location
- `Friction`: Slows movement proportionate to the particle's present velocity
- `Gravity`: Sets acceleration at an angle
- `Turbulence`: Liquid-like behavior based on a noise image
- `Wander`: Changes the route randomly
- `GroupGoal`: Changes the state of a particle group
- `SpriteGoal`: Changes the state of a sprite particle

Let's understand the use of `ParticleSystem` with the following example:

```
ParticleSystem {
    id: particleSystem
    anchors.fill: parent
    Image {
        source: "qrc:/logo.png"
        anchors.centerIn: parent
    }
    ImageParticle {
        system: particleSystem
        source: "qrc:/particle.png"
        colorVariation: 0.5
        color: "#00000000"
    }
    Emitter {
        id: emitter
        system: particleSystem
        enabled: true
        x: parent.width/2; y: parent.height/2
```

```
maximumEmitted: 8000; emitRate: 6000
size: 4 ; endSize: 24
sizeVariation: 4
acceleration: AngleDirection {
angleVariation: 360; magnitude: 360;
}
}
}
```

In the preceding code, we have used the Qt logo, which is emitting particles around it. We have created an instance of `ImageParticle` that creates particles that are emitted by `Emitter`. The `AngleDirection` type is used to decide the angle and direction of particle emission. Since we want the particles to be emitted around the logo, we have used 360 for both attributes. The output of the preceding example is shown in *Figure 8.8*:

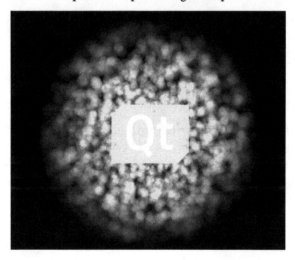

Figure 8.8 – Output of the above particle system example

You can explore more about these QML types on the following website:

`https://qmlbook.github.io/`

In this section, we discussed different types of drawing mechanisms and components in Qt Quick. In the next section, we will learn how to do animation in Qt Widgets.

Animation in Qt Widgets

The animation framework simplifies the process of animating a GUI element by allowing its properties to be animated. **Easing curves** are used to control the animations. Easing curves describe a function that controls the animation's speed, resulting in various acceleration and deceleration patterns. Qt supports linear, quadratic, cubic, quartic, sine, exponential, circular, and elastic easing curves. The property animation class provided by Qt, known as the QPropertyAnimation class, is one of the more common ways to animate a GUI element. This class is part of the animation framework, and it uses Qt's timer system to change the properties of a GUI element over a specified time period.

To create animations for our GUI application, Qt provides us with several subsystems, including a timer, timeline, animation framework, state machine framework, and the Graphics View framework.

Let's discuss how to use property animation with QPushButton in the following code:

```
QPropertyAnimation *animatateButtonA = new
QPropertyAnimation(ui->pushButtonA, "geometry");
animatateButtonA->setDuration(2000);
animatateButtonA->setStartValue(ui->pushButtonA->geometry());
animatateButtonA->setEndValue(QRect(100, 150, 200, 300));
```

In the preceding code snippet, we animated a push button from one position to another position and changed the button size. You can add easing curve to control the animation simply by adding it to the property animation before calling the start() function. You can also experiment with different types of easing curves to see which one works best for you.

Property animations and animation groups are both inherited from the QAbstractAnimator class. Hence, you can add one animation group to another to create a more complex, nested animation group. Qt currently provides two types of animation group classes, QParallelAnimationGroup and QSequentialAnimationGroup.

Let's use the QSequentialAnimationGroup group to manage the states of the animations within it:

```
QSequentialAnimationGroup *group = new
QSequentialAnimationGroup;
group->addAnimation(animatateButtonA);
```

```
group->addAnimation(animatateButtonB);
group->addAnimation(animatateButtonC);
```

You can explore more about Qt's animation framework at the following link:

`https://doc.qt.io/qt-6/animation-overview.html`

In this section, we discussed animation in Qt Widgets. In the next section, you will learn how to do animation in Qt Quick.

Animation and transitions in Qt Quick

In this section, you will learn how to create animation and add transitions in Qt Quick. To create an animation, you need to choose a proper animation type for the type of the property that is to be animated and then apply the animation for the required behavior.

Qt Quick has different types of animations, such as the following:

- `Animator`: It is a special type of animation that operates directly on Qt Quick's scene graph.

- `AnchorAnimation`: It is used for animating an anchor change.

- `ParallelAnimation`: It runs animations in parallel.

- `ParentAnimation`: It is used for animating a parent change.

- `PathAnimation`: It animates an item along a path.

- `PauseAnimation`: It enables pauses during animations.

- `PropertyAnimation`: It animates changes in property values.

- `SequentialAnimation`: It runs animations sequentially.

- `ScriptAction`: During an animation, it allows JavaScript to be executed.

- `PropertyAction`: It can change a property immediately during an animation, without the need to animate a property change.

Figure 8.9 shows the hierarchy of animation classes:

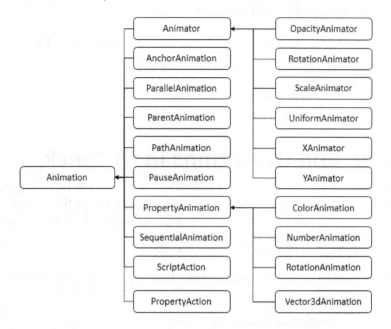

Figure 8.9 – The hierarchy of animation classes in Qt Quick

PropertyAnimation provides a way to animate changes to a property's value. Different subclasses of PropertyAnimation are as follows:

- ColorAnimation: Animates changes in color values
- NumberAnimation: Animates changes in qreal-type values
- RotationAnimation: Animates changes in rotation values
- Vector3dAnimation: Animates changes in QVector3d values

It can be used to define animations in several ways:

- In a Transition
- In a Behavior
- As a property
- In a signal handler
- Standalone

Property values are animated by applying animation types to them. To create smooth transitions, animation types will interpolate property values. State transitions can also assign animations to state changes:

- `SmoothedAnimation`: It is a specialized `NumberAnimation` subclass. In animation, when the target value is changed, `SmoothAnimation` ensures smooth changes.

- `SpringAnimation`: With its specialized attributes including mass, damping, and epsilon, it provides a spring-like animation.

Animation can be set for an object in different ways:

- Direct property animation

- Predefined targets and properties

- Animation as behaviors

- Transitions during state changes

Animations are created by applying animation objects to property values in order to change the properties gradually over time. Smooth movements are used in these property animations by interpolating values between property value changes. Property animations allow for different interpolations and timing controls via easing curves.

The following code snippet demonstrates two `PropertyAnimation` objects using predefined properties:

```
Rectangle {
    id: rect
    width: 100; height: 100
    color: "green"
    PropertyAnimation on x { to: 200 }
    PropertyAnimation on y { to: 200 }
}
```

In the preceding example, the animation will begin as soon as the `Rectangle` is loaded and is applied to its x and y values automatically. Here, we have used the `<AnimationType> on <Property>` syntax. Hence, it is not required to set the target and the property values to x and y.

Animations may be shown sequentially or parallelly. While sequential animations play a group of animations serially, parallel animations play a group of animations at the same time. Therefore, when animations are grouped inside a `SequentialAnimation` or a `ParallelAnimation`, they will be played sequentially or parallelly. `SequentialAnimation` can also be used for playing `Transition` animations since transition animations are automatically played in parallel. You can group the animations to ensure that all animations within a group are applied to the same property.

Let's use `SequentialAnimation` to animate the rectangle's `color` in the following example:

```
import QtQuick
import QtQuick.Window
Window {
    width: 640
    height: 480
    visible: true
    title: qsTr("Sequential Animation Demo")
    Rectangle {
        anchors.centerIn: parent
        width: 100; height: 100
        radius: 50
        color: "red"
        SequentialAnimation on color {
            ColorAnimation { to: "red"; duration: 1000 }
            ColorAnimation { to: "yellow"; duration: 1000 }
            ColorAnimation { to: "green"; duration: 1000 }
            running:true
            loops: Animation.Infinite
        }
    }
}
```

In the preceding example, we have used `SequentialAnimation` on the `color` property using the `<AnimationType> on <Property>` syntax. As a result, the child `ColorAnimation` objects are automatically added to this property, and no `target` or `property` animation values are needed.

You can use `Behavior` animations to set the default property animations. Animations specified in `Behavior` types are applied to the property and animate any property value changes. To intentionally enable or disable the behavior animations, you can use the `enabled` property. You can use several methods to assign behavior animations to properties. One of the methods is the `Behavior on <property>` declaration. It conveniently assigns a behavior animation onto a property.

`Animator` types are distinct from normal `Animation` types. Let's create a simple example where we rotate an image using an `Animator`:

```
import QtQuick
import QtQuick.Window
Window {
    width: 640
    height: 480
    visible: true
    title: qsTr("Animation Demo")
    Image {
        anchors.centerIn: parent
        source: "qrc:/logo.png"
        RotationAnimator on rotation {
            from: 0; to: 360;
            duration: 1000
            running:true
            loops: Animation.Infinite
        }
    }
}
```

In the preceding example, we have used the `RotationAnimator` type, which is used to animate the rotation of an `Image` QML type.

In this section, we discussed different types of animations in Qt Quick and created several examples. In the next section, we will discuss how to control animations.

Controlling animations

Controlling animations can be done in a variety of ways. The `Animation` type is the ancestor of all animation types. This type does not allow the creation of `Animation` objects. It equips a user with the necessary properties and methods to use animation types. All animation types consist of `start()`, `stop()`, `resume()`, `pause()`, `restart()`, and `complete()`, and they control how animations are executed.

The animation's interpolation between the start and end values is defined by the easing curves. Different easing curves may extend beyond the defined interpolation range. The easing curves make it easier to create animation effects such as bounce, acceleration, deceleration, and cyclical animations.

In a QML object, each property animation may have a distinct easing curve. The curve can be controlled with various parameters and some of these parameters are unique to a particular curve. Visit the easing documentation for more information on easing curves.

In this section, you learned about the way to control animations in Qt Quick. In the next section, you will learn how to use states and transitions.

States, state machine, and transitions in Qt Quick

Qt Quick states are property configurations in which a property's value can change to reflect different states. State changes cause abrupt changes in property; animations smooth transitions to create visually appealing state changes. Types for creating and executing state graphs in QML are provided by the Declarative State Machine Framework. Consider using the QML states and transitions for user interfaces with multiple visual states that are independent of the application's logical state.

You can import the state machine module and the QML types into your application by adding the following statement:

```
import QtQml.StateMachine
```

Please note that there are two ways to define the states in QML. One is provided by `QtQuick` and the other by the `QtQml.StateMachine` module.

> **Important note**
>
> While using QtQuick and QtQml.StateMachine in a single QML
> file, make sure to import QtQml.StateMachine after QtQuick. In
> this approach, the State type is provided by the Declarative State Machine
> Framework, not by QtQuick. To avoid any ambiguity with QtQuick's State
> item, you can import QtQml.StateMachine into a different namespace.

To interpolate property changes caused by state changes, the Transition type can
include animation types. Bind the transition to the transitions property to assign
it to an object.

A button can have two states: pressed and released. For each state, we can assign
a different property configuration. A transition would animate the transition from
pressed to released. Similarly, there would be animation when switching from
the released to the pressed state.

Let's have a look at the following example.

Create a circular LED using the Rectangle QML type and add a MouseArea to it.
Assign the default state as OFF and the color as green. On mouse press, we want
to change the LED color to red and once the mouse is released, the LED becomes
green again:

```
Rectangle {
    id:led
    anchors.centerIn: parent
    width: 100
    height: 100
    radius: 50
    color: "green"
    state: "OFF"
    MouseArea {
        anchors.fill: parent
        onPressed: led.state = "ON"
        onReleased: led.state = "OFF"
    }
}
```

Next, define the states. In this example, we have two states, ON and OFF. Here, we are manipulating the color property based on the state change:

```
states: [
      State {
            name: "ON"
            PropertyChanges { target: led; color: "red"}
      },
      State {
            name: "OFF"
            PropertyChanges { target: led; color: "green"}
      }
   ]
```

You can add an animation to the transitions. Let's add ColorAnimation to the transition to make it smooth and attractive:

```
transitions: [
   Transition {
         from: "ON"
         to: "OFF"
         ColorAnimation { target: led; duration: 100}
      },
      Transition {
         from: "OFF"
         to: "ON"
         ColorAnimation { target: led; duration: 100}
      }
   ]
```

In the preceding example, we have used two states, ON and OFF. We have used MouseArea to change the states based on mouse press and release events. When the state is ON, the rectangle color changes to red, and when it is OFF, the color changes to green. Here, we have also used Transition to switch between the states.

When the `to` and `from` properties are bound to the state's name, the transition will be associated with the state change. For simple or symmetric transitions, setting the `to` property to the wild card symbol `"*"` implies that the transition applies to any state change:

```
transitions: Transition {
    to: "*"
    ColorAnimation { target: led; duration: 100 }
}
```

You can explore more about the State Machine QML API at the following link:

`https://doc.qt.io/qt-6/qmlstatemachine-qml-guide.html`

In this section, you learned about the state machine in Qt Quick. In the next section, you will learn how to use the state machine in Qt Widgets.

The state machine in Qt Widgets

Classes in the State Machine framework are available for creating and executing state graphs. The State Machine framework provides an API and execution model for effectively embedding state chart elements and semantics in Qt applications. The framework is tightly integrated with Qt's meta-object system.

There was a major change to the State Machine framework in Qt 6. The APIs were missing from the Qt 6.0.x core module. With Qt 6.1, the module was restored as the **statemachine** module. So, you won't be able to run it in Qt 6.0.x versions and you will have to add `statemachine` to the `.pro` file to use the framework.

If you are using a `qmake` based build system, then add the following line to your `.pro` file:

```
QT += statemachine
```

If you are using a *CMake* based build system, then add the following to `CMakeLists.txt`:

```
find_package(Qt6 COMPONENTS StateMachine REQUIRED)
target_link_libraries(mytarget PRIVATE Qt6::StateMachine)
```

You will need the following headers inside your C++ source file:

```
#include <QStateMachine>
#include <QState>
```

Let's create a simple Qt Widgets application that implements the state-machine. Modify the UI form by adding `QLabel` and `QPushButton`:

1. Add the following code to the constructor of your custom C++ class:

```
QState *green = new QState();
green->assignProperty(ui->pushButton, "text", "Green");
green->assignProperty(ui->led,
"styleSheet","background-color: rgb(0, 190, 0);");
green->setObjectName("GREEN");
```

2. In the preceding code, we created a state to show the green-colored LED. Next, we will create another state for the red-colored LED:

```
QState *red = new QState();
red->setObjectName("RED");
red->assignProperty(ui->pushButton, "text", "Red");
red->assignProperty(ui->led, "styleSheet", "background-
color: rgb(255, 0, 0);");
```

3. Add transitions for the state change events when the button is toggled:

```
green->addTransition(ui->pushButton,
&QAbstractButton::clicked,red);
red->addTransition(ui->pushButton,
&QAbstractButton::clicked,green);
```

4. Now create a state machine instance and add the states to it:

```
QStateMachine *machine = new QStateMachine(this);
machine->addState(green);
machine->addState(red);
machine->setInitialState(green);
```

5. The last step is to start the state machine:

```
machine->start();
```

6. When you run the previous example, you will see an output window like the following:

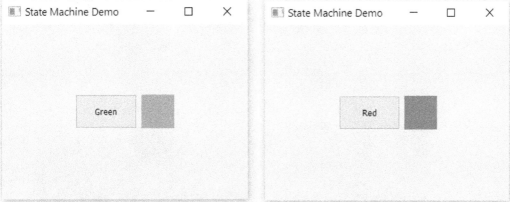

Figure 8.10 – Output of the application using the state machine in Qt Widgets

The preceding diagram reinforces that in a parent state machine, only the states of the child state machine can be specified as transition targets. States of the parent state machine, on the other hand, cannot be specified as targets of transitions in the child state machine.

The following article nicely captures the performance considerations while using a state machine:

`https://www.embedded.com/how-to-ensure-the-best-qt-state-machine-performance/`

In this section, we learned about state machines and their usage in Qt Widgets. We discussed how to implement state machines in both Qt Widgets and Qt Quick. Let's summarize what we learned in this chapter.

Summary

In this chapter, we discussed different graphics APIs and we learned how to use the `QPainter` class to draw graphics both on and off the screen. We also looked into the Graphics View framework and scene graph rendering mechanism. We saw how Qt provides the `QPaintDevice` interface and the `QPainter` class to perform graphics operations throughout this chapter. We also discussed the Graphics View classes, OpenGL framework, and shader tools. At the end of the chapter, we explored the animation and state machine framework in both Qt Widgets and Qt Quick.

In *Chapter 9, Testing and Debugging*, we will learn about debugging and testing in Qt. It will help you to find the root cause of issues and fix defects.

9
Testing and Debugging

Debugging and testing are essential parts of software development. In this chapter, you will learn how to debug Qt projects, about different debugging techniques, and about debuggers supported by Qt. Debugging is the process of discovering the root cause of an error or undesired behavior and resolving it. We will also discuss unit testing using the Qt Test framework. Qt Test is a unit testing framework for Qt-based applications and libraries. It has all of the features that most unit testing frameworks provide. Additionally, it provides support for testing **Graphical User Interfaces (GUIs)**. This module helps in writing unit tests for Qt-based applications and libraries in a convenient way. You will also learn techniques to test a GUI using different GUI testing tools.

Specifically, we will discuss the following topics:

- Debugging in Qt
- Debugging strategies
- Debugging a C++ application
- Debugging a Qt Quick application
- Testing in Qt

- Integrating with Google's C++ testing framework
- Testing Qt Quick applications
- GUI testing tools

By the end of this chapter, you will be familiar with debugging and testing techniques for your Qt application.

Technical requirements

The technical requirements for this chapter include minimum versions of Qt 6.0.0 and Qt Creator 4.14.0 installed on the latest version of a desktop platform such as Windows 10, Ubuntu 20.04, or macOS 10.14.

All the code used in this chapter can be downloaded from the following GitHub link: `https://github.com/PacktPublishing/Cross-Platform-Development-with-Qt-6-and-Modern-Cpp/tree/master/Chapter09`.

> **Important note**
> The screenshots used in this chapter are taken from the Windows platform.
> You will see similar screens based on the underlying platform on your machine.

Debugging in Qt

In software development, technical problems arise often. To address these issues, we must first identify and resolve all of them before releasing our application to the public to maintain quality and our reputation. Debugging is a technique for locating these underlying technological issues.

In the coming sections, we will discuss popular debugging techniques used by software engineers to ensure their software's stability and quality.

Debuggers supported by Qt

Qt supports several different types of debuggers. The debugger you use can vary depending on the platform and compiler you're using for your project. The following is a list of debuggers that are widely used with Qt:

- **GNU Symbolic Debugger** (**GDB**) is a cross-platform debugger developed by the GNU Project.

- **Microsoft Console Debugger (CDB)** is a debugger from Microsoft for Windows.

- **Low Level Virtual Machine Debugger (LLDB)** is a cross-platform debugger developed by the LLVM Developer group.

- **QML/JavaScript Debugger** is a QML and JavaScript debugger provided by the Qt company.

If you're using the MinGW compiler on Windows, you won't need to do any manual setup with GDB because it's typically included with your Qt installation. If you're using a different operating system, such as Linux, you may need to manually install it before linking it to Qt Creator. Qt Creator automatically detects the presence of the GDB and adds it to its debugger list.

You can also use **Valgrind** to debug your application. You can activate the Valgrind gdbserver by specifying either --vgdb=yes or --vgdb=full. You can specify --vgdb-error=number to activate gdbserver after a certain number of errors are displayed. If you set the value to 0, then gdbserver will be active at initialization, allowing you to set breakpoints before the application launches. It's worth noting that vgdb is included in the **Valgrind** distribution. It does not need to be installed separately.

If your favorite platform is Windows, you can install CDB on your machine. By default, the built-in debugger of Visual Studio won't be available. Therefore, you must install the CDB debugger separately by choosing debugging tools for Windows as an optional component when installing the Windows SDK. Qt Creator usually recognizes the existence of CDB and adds it to the debugger list under **Options**.

Android debugging is a little more challenging than debugging on a regular desktop environment. Different packages, such as JDK, Android SDK, and Android NDK, are required for Android development. On the desktop platform, you will need the **Android Debug Bridge (ADB)** driver to allow USB debugging. You must enable developer mode and accept USB debugging on the Android device to proceed.

The debugger used on macOS and iOS is **LLDB**. It is included with Xcode by default. Qt Creator will automatically detect its presence and link it with a kit. If you're familiar with debuggers and know what you're doing, you can also add non-GDB debuggers to your favorite IDE.

The debugger plugin determines a suitable native debugger for each package based on what's available on your machine. You can overcome this preference by adding new debuggers. You can find the available debuggers in the **Debuggers** tab present under the **Kits** settings under the **Options** menu as shown in *Figure 9.1*:

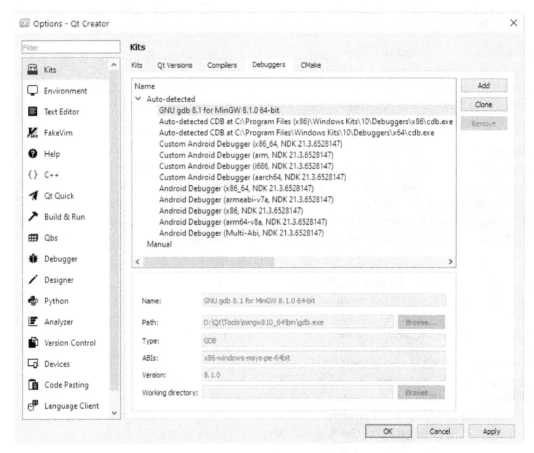

Figure 9.1 – The Debuggers tab under the Kits selection screen showing the Add button

In the **Debuggers** tab, you can see **Add**, **Clone**, and **Remove** buttons on the right side. You can clone an existing debugger configuration and modify it to suit your requirements. Alternatively, if you are aware of the debugger's details and configuration, then you can create a new debugger configuration using the **Add** button. You can also remove a faulty or obsolete debugger configuration by clicking the **Remove** button. Don't forget to click the **Apply** button to save your changes. Please note that you can't modify auto-detected debugger configurations.

In this section, we learned about various supported debuggers. In the next section, we will discuss how to debug an application.

Debugging strategies

There are different debugging strategies to find the root cause of an issue. Before attempting to locate a bug in the application, it is critical to thoroughly understand the program or library. You can't find mistakes if you don't know what you're doing. Only if you have a thorough understanding of the system and how it operates will you be able to identify bugs in the application. Previous experience can aid in the detection of similar types of bugs as well as the resolution of bugs. The individual expert's knowledge determines how easily the developer can locate the bug. You can add debug print statements and breakpoints to analyze the flow of the program. You can do forward analysis or backward analysis to track the bug's location.

When debugging, the following steps are used to find the root cause and resolve it:

1. Identify the issue.

2. Locate the issue.

3. Analyze the issue.

4. Resolve the issue.

5. Fix the side effects.

Regardless of the programming language or platform, the most important thing to know when debugging your application is which section of your code is causing the problem. You can find the faulty code in a number of ways.

If the defect is raised by your QA team or a user, then inquire when the issue occurred. Look at the log files or any error messages. Comment out the suspected section of the code, then build and run the application again to see if the issue persists. If the issue is reproducible, do forward and backward analysis by printing messages and commenting out lines of code before you find the one that's causing the issue.

You can also set a breakpoint in the built-in debugger to search for variable changes within your targeted feature. If one of the variables has updated to an unexpected value or an object pointer has become an invalid pointer, then you can easily identify it. Inspect all of the modules you used in the installer and ensure that you and your users have the same version number of the application. If you are using a different version or different branch, then check out the branch with the specified version tag, then debug the code.

In the next section, we will discuss how to debug your C++ code by printing debug messages and adding breakpoints.

Debugging a C++ application

The `QDebug` class can be used to print the value of a variable to the application output window. `QDebug` is similar to `std::cout` in the standard library, but it has the benefit of being part of Qt, which means it supports Qt classes out of the box and can display its value without the need for conversion.

To enable debugging messages, we must include the `QDebug` header as follows:

```
#include <QDebug>
```

Qt provides several global macros for generating different types of debug messages. They can be used for different purposes, mentioned as follows:

- `qDebug()` provides a custom debug message.
- `qInfo()` provides informational messages.
- `qWarning()` reports warnings and recoverable errors.
- `qCritical()` provides critical error messages and reports system errors.
- `qFatal()` provides fatal error messages before exiting.

You can see if your feature is working correctly by using `qDebug()`. After you've finished looking for the error, remove the line of code that contains `qDebug()` to avoid unwanted console logs. Let's look at how to use `qDebug()` to print out variables to the output pane with an example. Create a sample `QWidget` application and add a function, `setValue(int value)`, and add the following code inside the function definition:

```
int value = 500;
qDebug() << "The value is : " << value;
```

The preceding code will show the following output in the output window present at the bottom of Qt Creator:

```
The value is : 500
```

You can figure out whether the value was changed by another function by looking at how many times the function is used and called inside the application. If the debug message is printed multiple times, then it is invoked from multiple places. Check if the correct value is sent to all calling functions. To eliminate unnecessary console logs in the output console window, remove the line of code that contains `qDebug()` once you have finished looking for the issue. Alternatively, you may implement conditional compilation.

Let's look further into debugging and debugging options in Qt Creator:

1. You can see a **Debug** menu in the menu bar. When you click on it, you will see a context menu with submenus as shown in *Figure 9.2*:

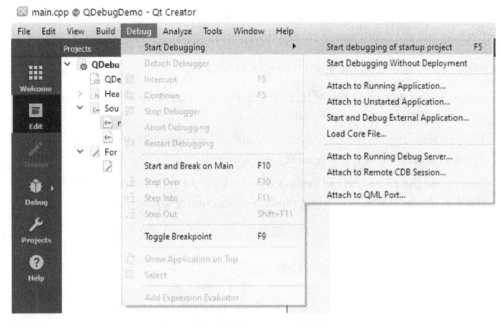

Figure 9.2 – Debug menu in Qt Creator

2. To start debugging, press *F5* or click on the start **Debug** button at the bottom left of Qt Creator as shown here:

Figure 9.3 – The Start debugging button in Qt Creator

3. If Qt Creator complains about the debugger with an error message, then check to see if your project package has a debugger.

4. If the error persists, close Qt Creator and go to your project folder, where you can delete the `.pro.user` file.

5. Then reload the project in Qt Creator. Your project will be reconfigured by Qt Creator, and the debug mode should now be functional.

A great way to debug your application is to set a breakpoint:

1. You will see a pop-up menu of three choices when you right-click on the line number of your script in Qt Creator.

2. You can also click on the line number to add a breakpoint. Click on the line number to set a breakpoint. You will see a red dot appearing on the line number.

3. Next, press the *F5* key on the keyboard or click on the start **Debug** button. Once you run the application in debug mode, you will notice a yellow arrow appearing on top of the first red dot:

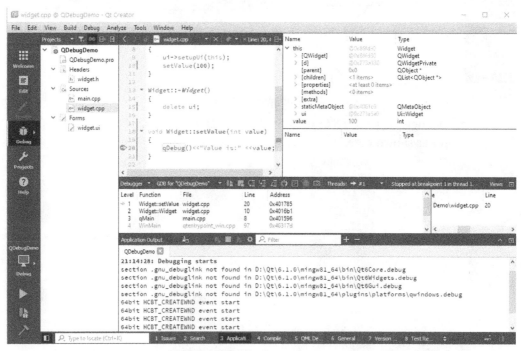

Figure 9.4 – Qt Creator showing debugging windows and breakpoints

4. The debugger has come to a halt at the first breakpoint. The variable, along with its meaning and type, will now be displayed in the **Locals** and **Expression** windows on the right-hand side of your Qt Creator.

5. This approach can be used to quickly examine the application. To remove a breakpoint, just click on the red dot icon once more or from the right-click context menu:

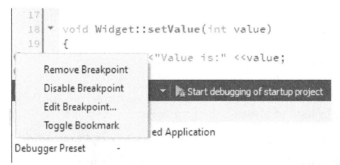

Figure 9.5 – Context menu showing right-click options on a breakpoint marking

It's important to remember that you must run your application in debug mode. This is because when you compile in debug mode, your application or library will have additional debugging symbols that allow your debugger to access information from the binary's source code, such as the names of identifiers, variables, and functions. This is the reason the application or library binaries are larger in file size when compiled in debug mode.

You can learn about more features and their usage in the following documentation:

`https://doc.qt.io/qt-6/debug.html`

> **Important note**
> Some anti virus applications prevent debuggers from retrieving information. One such anti virus is Avira. If it is installed on a production PC, the launching of the debugger could fail on the Windows platform.

In the next section, we will discuss how to debug a Qt Quick application and locate issues inside a QML file.

Debugging a Qt Quick application

In the last section, we discussed how to debug your C++ code. But you are probably still wondering how to debug code written in QML. Qt also has a provision to debug your QML code. When you are developing a Qt Quick application, there are a lot of options to troubleshoot issues. In this section, we will discuss various debugging techniques related to QML and how to use them.

Just like the QDebug class, there are different console APIs that are available for debugging in QML. They are as follows:

- Log: This is used to print general messages.
- Assert: This is used to verify an expression.
- Timer: This is used to measure the time spent between calls.
- Trace: This is used to print a stack trace of a JavaScript execution.
- Count: This is used to find the number of calls made to a function.
- Profile: This is used to profile QML and JavaScript code.
- Exception: It is used to print error messages.

The Console API provides several convenient functions to print different types of debug messages such as console.log(), console.debug(), console.info(), console.warn(), and console.error(). You can print a message with the value of a parameter as follows:

```
console.log("Value is:", value)
```

You can also check the creation of a component by adding the message inside Components.onCompleted: {...}:

```
Components.onCompleted: {
    console.log("Component created")
}
```

To verify that an expression is true, you can use console.assert(), such as the following, for example:

```
console.assert(value == 100, "Reached the maximum limit");
```

You will find the time spent between calls is logged by console.time() and console.timeEnd(). The stack trace of the JavaScript execution at the stage where it was called is printed by console.trace(). The function name, filename, line number, and column number are all included in the stack trace details.

`console.count()` returns the current number of times a piece of code has been executed, as well as a message. The QML and JavaScript profiling are activated when you use `console.profile()` and deactivated when `console.profileEnd()` is called. You can use `console.exception()` to print an error message along with the stack trace of the JavaScript execution.

You can add a breakpoint in the same way we discussed in an earlier section, as follows:

- To step into the code in the stack, click on the **Step Into** button on the toolbar or press *F11*.

- To step out, press *Shift + F11*. To hit the breakpoint, add a breakpoint at the end of the method and click **Continue**.

- Open the QML debugger console output pane to run JavaScript commands in the current context.

You can find the issues and watch the values while running your Qt Quick application. It will help you to find the portion of the code that is causing unexpected behavior and requires modification.

In this section, we learned about debugging in a QML environment. In the next section, we will discuss the testing framework in Qt.

Testing in Qt

Unit testing is a way of testing a simple application, class, or function using an automated tool. We will discuss what it is and why we would like to do it before going over how to incorporate it into our approach using Qt Test. Unit testing is the process of breaking down an application into its smallest functional units and then testing each unit with real-world situations within the initiative's framework. A unit is the smallest component of an application that can be tested. A unit test in procedural programming usually focuses on a function or process.

A unit in object-oriented programming is usually an interface, a class, or a single function. Unit testing identifies issues early in the implementation process. This covers glitches in the programmer's implementation as well as defects in or incomplete portions of the unit's specification. During the creation process, a unit test is a short code fragment developed by the developer of the unit to be tested. There are many unit testing tools to test your C++ code. Let's explore the benefits and features of Qt's testing framework.

Unit testing in Qt

Qt Test is a unit testing platform for Qt-based applications and libraries. Qt Test includes all of the features present in traditional unit testing applications, as well as plugins for testing graphical user interfaces. It helps make writing unit tests for Qt-based programs and libraries even easier. *Figure 9.6* shows the **Testing** section under **Options**:

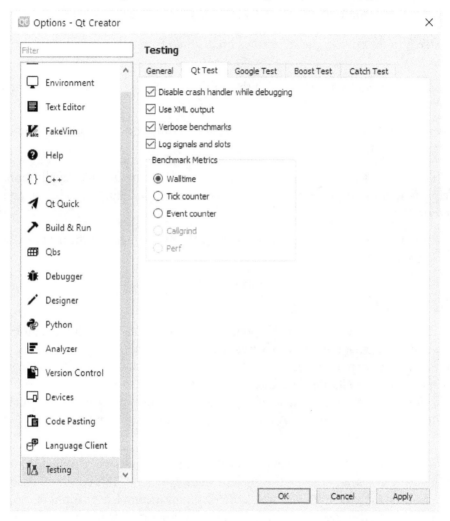

Figure 9.6 – Screenshot showing Qt Test preferences under the Qt Creator Options menu

Previously, unit testing may have been done manually, especially for GUI testing, but now there is a tool that allows you to write code to validate code automatically, which might seem counterintuitive at first, but it works properly. Qt Test is a specialized testing framework for unit testing based on Qt.

You have to add `testlib` in your project file (`.pro`) to use Qt's built-in unit testing module:

```
QT += core testlib
```

Next, run qmake to add the module available for your project. In order for the test system to find and implement it, you must use the `QTest` header and declare the test functions as private slots. The `QTest` header contains all functions and statements related to Qt Test. To use the `QTest` features, simply add the following line to your C++ file:

```
#include <QTest>
```

You should write test cases for every possible scenario, and then run the tests every time your baseline code changes to ensure that the system continues to behave as intended. It is an extremely useful tool for ensuring that any programming updates made don't break existing features.

Let's create a simple test application using Qt Creator's built-in wizard. Select **Auto Test Project** from the **New Project** menu as shown in *Figure 9.7*:

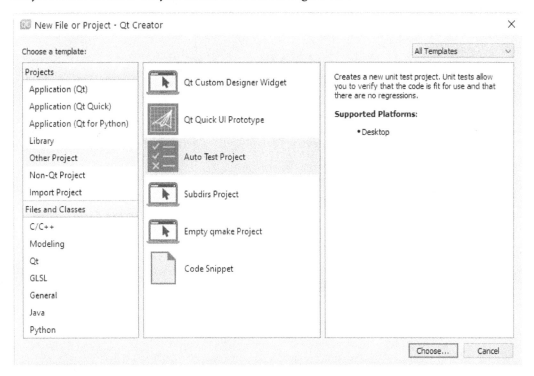

Figure 9.7 – New auto test project option in the project wizard

Once the test project skeleton is generated, you can modify the generated files to suit your needs. Open the `.pro` file of your test project and add the following lines of code:

```
QT += testlib
QT -= gui
CONFIG += qt console warn_on depend_includepath testcase
CONFIG -= app_bundle
TEMPLATE = app
SOURCES +=  tst_testclass.cpp
```

Let's create a C++ class named `TestClass`. We will add our test functions to this class. This class must be derived from `QObject`. Let's have a look at `tst_testclass.cpp`:

```cpp
#include <QtTest>
class TestClass : public QObject
{
    Q_OBJECT
public:
    TestClass() {}
    ~TestClass(){}
private slots:
    void initTestCase(){}
    void cleanupTestCase() {}
    void test_compareStrings();
    void test_compareValues();
};
```

In the preceding code, we have declared two test functions to test sample strings and values. You need to implement the test functions with a test scenario for the declared test cases. Let's compare two strings and do a simple arithmetic operation. You can use macros such as `QCOMPARE` and `QVERIFY` to test the values:

```cpp
void TestClass::test_compareStrings()
{
    QString string1 = QLatin1String("Apple");
    QString string2 = QLatin1String("Orange");
    QCOMPARE(string1.localeAwareCompare(string2), 0);
}
void TestClass::test_compareValues()
```

```
{
    int a = 10;
    int b = 20;
    int result = a + b;
    QCOMPARE(result,30);
}
```

To execute all the test cases, you have to add macros such as QTEST_MAIN() at the bottom of the file. The QTEST_MAIN() macro expands to a simple main() method that runs all the test functions. The QTEST_APPLESS_MAIN() macro is useful for simple standalone non-GUI tests where the QApplication object is not used. Use QTEST_GUILESS_MAIN() if the GUI is not required but an event loop is required:

```
QTEST_APPLESS_MAIN(TestClass)
#include "tst_testclass.moc"
```

To make the test case a standalone executable, we have added the QTEST_APPLESS_MAIN() macro and the moc generated file for the class. You may use a number of other macros to test the application. For further information, please visit the following link:

http://doc.qt.io/qt-6/qtest.html#macros

When you run the preceding example, you will see the output with the test results as shown here:

```
********* Start testing of TestClass *********
Config: Using QtTest library 6.1.0, Qt 6.1.0 (x86_64-little_
endian-llp64 shared (dynamic) release build; by GCC 8.1.0),
windows 10
64bit HCBT_CREATEWND event start
PASS   : TestClass::initTestCase()
FAIL!  : TestClass::test_compareStrings() Compared values are
not the same
   Actual   (string1.localeAwareCompare(string2)): -1
   Expected (0)                                   : 0
..\TestProject\tst_testclass.cpp(26) : failure location
PASS   : TestClass::test_compareValues()
PASS   : TestClass::cleanupTestCase()
Totals: 3 passed, 1 failed, 0 skipped, 0 blacklisted, 7ms
********* Finished testing of TestClass *********
```

You can see that one test case failed as it did not meet the test criteria. Similarly, you can add more test cases and fetch parameters from another class to test the functionality. You can also run all tests with the **Run All Tests** option from the **Tests** context menu from the Qt Creator menu bar as shown in *Figure 9.8*:

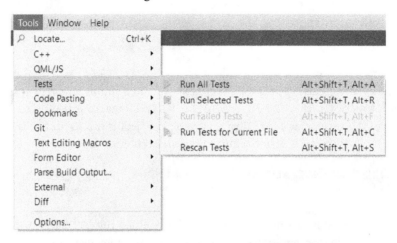

Figure 9.8 – Tests option under the Tools menu

You can also view all test cases in the left side project explorer view. Select **Tests** from the project explorer dropdown. You can enable or disable certain test cases in this window. *Figure 9.9* displays the two test cases we wrote earlier. You can also see that we are not using other test frameworks for this test project:

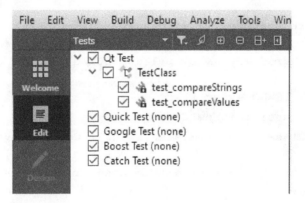

Figure 9.9 – Tests explorer option in the project explorer dropdown

You can use several `QTest` convenient functions to simulate GUI events such as keyboard or mouse events. Let's look at their usage with a simple code snippet:

```
QTest::keyClicks(testLineEdit, "Enter");
QCOMPARE(testLineEdit->text(), QString("Enter"));
```

In the preceding code, the test code simulates a keyboard text `Enter` event on a `lineedit` control and then verifies the entered text. You can also simulate mouse-click events using `QTest::mouseClick()`. You can use it as follows:

```
QTest::mouseClick(testPushBtn, Qt::LeftButton);
```

Qt's Test framework is also useful in **test-driven development** (TDD). In TDD, you write a test first, then code the actual logic. The test will initially fail as there is no implementation. You then write the bare minimum code required to pass the test before moving on to the next test. This is how you iteratively develop a feature before you have implemented the necessary functionality.

In this section, we learned how to create test cases and simulate GUI interaction events. In the next section, you will learn how to use Google's C++ testing framework.

Integrating with Google's C++ testing framework

GoogleTest is a testing and mocking framework developed by Google. The **GoogleMock** project has been merged into GoogleTest. GoogleTest requires a compiler that supports at least C++11 standards. It is a cross-platform test framework and it supports major desktop platforms such as Windows, Linux, and macOS. It helps you write better C++ tests with advanced features such as mocking. You can integrate Qt Test with GoogleTest to get the best of both frameworks. If you intend to use both testing framework features, then you should use GoogleTest as the primary testing framework and inside the test cases, you can use Qt Test's features.

Qt Creator has built-in support for GoogleTest. You can find the **Google Test** tab in the **Testing** section on the **Options** screen and set your global GoogleTest preferences as shown in *Figure 9.10*:

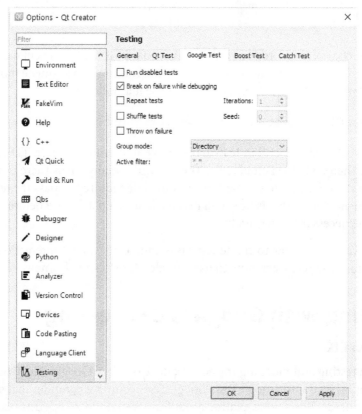

Figure 9.10 – The Google Test tab in the Testing section under the Options menu

You can download the GoogleTest source code from the following link:

`https://github.com/google/googletest`

You can learn more about features and their usage in the following documentation:

`https://google.github.io/googletest/primer.html`

After you download the source code, build the libraries before creating a sample application. You can also build the unified GoogleTest source code along with your test project. Once you generate the libraries, follow these steps to run your GoogleTest application:

1. To create a simple GoogleTest application using Qt Creator's built-in wizard, select **Auto Test Project** from the **New Project** menu. Then follow through the screens until you come across **Project and Test Information**.

2. On the **Project and Test Information** screen, select **Google Test** for **Test framework**. Then add information for the **Test suite name** and **Test case name** fields as shown in *Figure 9.11*:

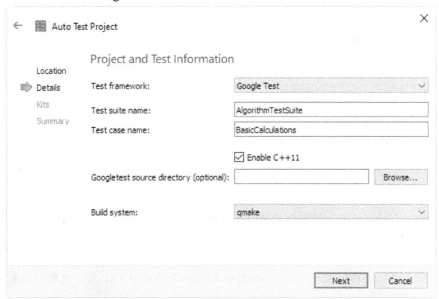

Figure 9.11 – Google Test option in the project creation wizard

3. In the next step, you can fill in the **Googletest source directory** field or you can add it later by editing the `.pro` file.

Figure 9.12 – Option to add the GoogleTest source directory in the project creation wizard

4. Click **Next** and follow the instructions to generate the skeleton of the project.

5. To use GoogleTest, you have to add the header file into your test project:

```
#include "gtest/gtest.h"
```

6. You can see the main function has already been created by the wizard:

```
#include "tst_calculations.h"
#include "gtest/gtest.h"
int main(int argc,char *argv[])
{
    ::testing::InitGoogleTest(&argc,argv);
    return RUN_ALL_TESTS();
}
```

7. You can create a simple test case with the following syntax:

```
TEST(TestCaseName, TestName) { //test logic }
```

8. GoogleTest also provides macros such as `ASSERT_*` and `EXPECT_*` to check conditions and values:

```
ASSERT_TRUE(condition)
ASSERT_EQ(expected,actual)
ASSERT_FLOAT_EQ(expected,actual)
EXPECT_DOUBLE_EQ (expected, actual)
```

In most cases, it is a standard procedure to do some custom initialization work before running multiple tests. If you want to evaluate a test's time/memory footprint, you'll have to write some test-specific code. Test fixtures help in setting up specific testing requirements. The `fixture` class is derived from the `::testing::Test` class. Please note that instead of `TEST`, the `TEST_F` macro is used. You can allocate resources and do initializations in the constructor or in the `SetUp()` function. Similarly, you can deallocate in the destructor or in the `TearDown()` function. A test function inside a text fixture is defined as follows:

```
TEST_F(TestFixtureName, TestName) { //test logic }
```

9. To create and use a test fixture, create a class derived from the ::testing::Test class as follows:

```cpp
class PushButtonTests: public ::testing::Test
{
protected:
    virtual void SetUp()
    {
        pushButton = new MyPushButton(0);
        pushButton ->setText("My button");
    }
};
TEST_F(PushButtonTests, sizeConstraints)
{
    EXPECT_EQ(40, pushButton->height());
    EXPECT_EQ(200, pushButton->width());
    pushButton->resize(300,300);
    EXPECT_EQ(40, pushButton->height());
    EXPECT_EQ(200, pushButton->width());
}
TEST_F(PushButtonTests, enterKeyPressed)
{
    QSignalSpy spy(pushButton, SIGNAL(clicked()));
    QTest::keyClick(pushButton, Qt::Key_Enter);
    EXPECT_EQ(spy.count(), 1);
}
```

In the preceding code, we created a custom push button inside the SetUp() function. Then we tested two test functions to test the size and *Enter* key handling.

10. When you run the preceding test, you will see the test results in the output window.

GoogleTest builds a new test fixture at runtime for each test specified with TEST_F(). It instantly initializes by calling the SetUp() function and runs the test. Then it calls TearDown() to do the cleanup, and removes the test fixture. It is important to note that different tests within the same test suite can have different test fixture objects. Before building the next test fixture, GoogleTest always deletes the previous one. It does not reuse test fixtures for multiple tests. Any modifications done to the fixture by one test have no effect on the other tests.

We discussed how to create a GoogleTest project with a simple test case and how to design a test fixture or test suite. Now you can create test cases for your existing C++ application. GoogleTest is a very mature test framework. It also integrates the mocking mechanism that was earlier available under GoogleMock. Explore different features and experiment with test cases.

There is also a ready-made GUI tool that integrates both test frameworks to test your Qt application. **GTest Runner** is a Qt-based automated test runner and GUI with powerful features for Windows and Linux platforms. However, the code is not actively maintained and is not upgraded to Qt 6. You can learn more about features and usages of GTest Runner at the following link:

`https://github.com/nholthaus/gtest-runner`

In this section, you learned how to use `QTest` and `GoogleTest` together. You have seen the features of both testing frameworks. You can create mock objects using the GoogleMock feature of the GoogleTest framework. Now you can write your own test fixtures for a custom C++ class or custom widget. In the next section, we will discuss testing in Qt Quick.

Testing Qt Quick applications

Qt Quick Test is a framework created for the unit testing of Qt Quick applications. Test cases are written in JavaScript and use the `TestCase` QML type. Functions with names beginning with `test_` are identified as test cases that need to be executed. The test harness recursively searches for the required source directory for `tst_ *.qml` files. You can keep all test `.qml` files under one directory and define the `QUICK_TEST_SOURCE_DIR`. If it is not defined, then only `.qml` files available in the current directory will be included during test execution. Qt doesn't ensure binary compatibility for the Qt Quick Test module. You have to use the appropriate version of the module.

You have to add `QUICK_TEST_MAIN()` to the C++ file to begin the execution of the test cases, as shown next:

```
#include <QtQuickTest>
QUICK_TEST_MAIN(testqml)
```

You need to add the `qmltest` module to enable Qt Quick Test. Add the following lines of code to the `.pro` file:

```
QT += qmltest
TEMPLATE = app
TARGET = tst_calculations
CONFIG += qmltestcase
SOURCES += testqml.cpp
```

Let's see a demo of a basic arithmetic calculation to see how the module works. We will do some calculations such as addition, subtraction, and multiplication and intentionally make some mistakes so that test cases will fail:

```
import QtQuick
import QtTest
TestCase {
    name: "Logic Tests"
    function test_addition() {
        compare(4 + 4, 8, "Logic: 4 + 4 = 8")
    }
    function test_subtraction() {
        compare(9 - 5, 4, "Logic: 9 - 5 = 4")
    }
    function test_multiplication() {
        compare(3 * 3, 6, "Logic: 3 * 3 = 6")
    }
}
```

When you run the preceding example, you will see the output with the test results as follows:

```
********* Start testing of testqml *********
Config: Using QtTest library 6.1.0, Qt 6.1.0 (x86_64-little_
endian-llp64 shared (dynamic) release build; by GCC 8.1.0),
windows 10
PASS    : testqml::Logic Tests::initTestCase()
PASS    : testqml::Logic Tests::test_addition()
FAIL!   : testqml::Logic Tests::test_multiplication()Logic: 3 *
3 = 6
   Actual   (): 9
   Expected (): 6
C:\Qt6Book\Chapter09\QMLTestDemo\tst_calculations.qml(15) :
failure location
PASS    : testqml::Logic Tests::test_subtraction()
PASS    : testqml::Logic Tests::cleanupTestCase()
Totals: 4 passed, 1 failed, 0 skipped, 0 blacklisted, 3ms
********* Finished testing of testqml *********
```

Please note that `cleanupTestCase()` is called right after the test execution has been completed. This function can be used to clean up before everything is destructed.

You can also perform data-driven tests as shown here:

```qml
import QtQuick
import QtTest
TestCase {
    name: "DataDrivenTests"
    function test_table_data() {
        return [
            {tag: "10 + 20 = 30", a: 10, b: 20, result: 30
},
            {tag: "30 + 60 = 90", a: 30, b: 60, result: 90
},
            {tag: "50 + 50 = 100", a: 50, b: 50, result: 50
},
        ]
    }
    function test_table(data) {
        compare(data.a + data.b, data.result)
    }
}
```

Please note that the table data can be provided to a test using a function name that ends with _data. When you run the preceding example, you will see the output with the test results as follows:

```
********* Start testing of main *********
Config: Using QtTest library 6.1.0, Qt 6.1.0 (x86_64-little_
endian-llp64 shared (dynamic) release build; by GCC 8.1.0),
windows 10
PASS    : main::DataDrivenTests::initTestCase()
PASS    : main::DataDrivenTests::test_table(10 + 20 = 30)
PASS    : main::DataDrivenTests::test_table(30 + 60 = 90)
FAIL!   : main::DataDrivenTests::test_table(50 + 50 = 100)
Compared values are not the same
   Actual   (): 100
   Expected (): 50
```

```
C:\Qt6Book\Chapter09\QMLDataDrivenTestDemo\tst_datadriventests.
qml(14) : failure location
PASS    : main::DataDrivenTests::cleanupTestCase()
Totals: 4 passed, 1 failed, 0 skipped, 0 blacklisted, 3ms
********* Finished testing of main *********
```

You can also run benchmark tests in QML. The Qt benchmark framework will run functions with names that begin with benchmark_ several times, with an average timing value recorded for the runs. It is similar to the QBENCHMARK macro in the C++ version of **QTestLib**. You can prefix the test function name with benchmark_once_ to get the effect of the QBENCHMARK_ONCE macro. Let's have a look at the following benchmarking example:

```qml
import QtQuick
import QtTest
TestCase {
    id: testObject
    name: "BenchmarkingMyItem"
    function benchmark_once_create_component() {
        var component = Qt.createComponent("MyItem.qml")
        var testObject = component.createObject(testObject)
        testObject.destroy()
        component.destroy()
    }
}
```

In the preceding example, we are creating a custom QML element. We want to measure how much time it takes to create the element. Hence, we wrote the preceding benchmark code. A normal benchmark test runs multiple times and shows the duration of the operation. Here, we have benchmarked the creation once. This technique is very useful in evaluating the performance of your QML code.

When you run the preceding example, you will see the output with the test results as follows:

```
********* Start testing of testqml *********
Config: Using QtTest library 6.1.0, Qt 6.1.0 (x86_64-little_
endian-llp64 shared (dynamic) release build; by GCC 8.1.0),
windows 10
PASS    : testqml::BenchmarkingMyItem::initTestCase()
```

```
PASS     : testqml::BenchmarkingMyItem::benchmark_once_create_
component()
PASS     : testqml::BenchmarkingMyItem::benchmark_once_create_
component()
RESULT : testqml::benchmark_once_create_component:
     0 msecs per iteration (total: 0, iterations: 1)
PASS     : testqml::BenchmarkingMyItem::cleanupTestCase()
QWARN    : testqml::UnknownTestFunc()
QQmlEngine::setContextForObject(): Object already has a
QQmlContext
Totals: 4 passed, 0 failed, 0 skipped, 0 blacklisted, 5ms
********* Finished testing of testqml *********
```

To run the benchmark multiple times, you can remove the `once` keyword from the test case as follows: `function benchmark_create_component() {...}`. You can also test dynamically created objects using `Qt.createQmlObject()`.

There is also a benchmarking tool named **qmlbench** for benchmarking the overall performance of a Qt application. It is a feature-rich benchmarking tool available under **qt-labs**. The tool also helps in measuring the refresh rate of the user interface. You can explore more about this tool at the following link:

`https://github.com/qt-labs/qmlbench`

Like a C++ implementation, you can also simulate keyboard events such as `keyPress()`, `keyRelease()`, and `keyClick()` in QML. The events are delivered to the QML object that is currently being focused on. Let's have a look at the following example:

```qml
import QtQuick
import QtTest
MouseArea {
    width: 100; height: 100
    TestCase {
        name: "TestRightKeyPress"
        when: windowShown
        function test_key_click() {
            keyClick(Qt.Key_Right)
        }
    }
}
```

In the preceding example, the keyboard event is delivered after the QML viewing window has been displayed. Attempts to deliver events before that will be unsuccessful. To keep track of when the window is shown, the when and windowShown properties are used.

When you run the preceding example, you will see the output with the test results as follows:

```
********* Start testing of testqml *********
Config: Using QtTest library 6.1.0, Qt 6.1.0 (x86_64-little_
endian-llp64 shared (dynamic) release build; by GCC 8.1.0),
windows 10
PASS    : testqml::TestRightKeyPress::initTestCase()
QWARN   : testqml::TestRightKeyPress::test_key_click()
QQmlEngine::setContextForObject(): Object already has a
QQmlContext
PASS    : testqml::TestRightKeyPress::test_key_click()
PASS    : testqml::TestRightKeyPress::cleanupTestCase()
Totals: 3 passed, 0 failed, 0 skipped, 0 blacklisted, 25ms
********* Finished testing of testqml *********
```

You can use SignalSpy to watch signal emission. In the following example, we have used SignalSpy to detect the clicked signal on a Button. When the signal is emitted, the clickSpy count is increased:

```
import QtQuick
import QtQuick.Controls
import QtTest
Button {
    id: pushButton
    SignalSpy {
        id: clickSpy
        target: pushButton
        signalName: "clicked"
    }
    TestCase {
        name: "PushButton"
        function test_click() {
            compare(clickSpy.count, 0)
            pushButton.clicked();
            compare(clickSpy.count, 1)
```

```
            }
        }
    }
```

When you run the preceding example, you will see the output with the test results as follows:

```
********* Start testing of testqml *********
Config: Using QtTest library 6.1.0, Qt 6.1.0 (x86_64-little_
endian-llp64 shared (dynamic) release build; by GCC 8.1.0),
windows 10
PASS    : testqml::PushButton::initTestCase()
PASS    : testqml::PushButton::test_click()
PASS    : testqml::PushButton::cleanupTestCase()
Totals: 3 passed, 0 failed, 0 skipped, 0 blacklisted, 5ms
********* Finished testing of testqml *********
```

The QUICK_TEST_MAIN_WITH_SETUP macro is used to execute C++ code before any of the QML tests are run. This can be useful for setting context properties on the QML engine. A test application can include several TestCase instances. The application terminates after running all test cases. You can enable or disable test cases from the **Tests** explorer:

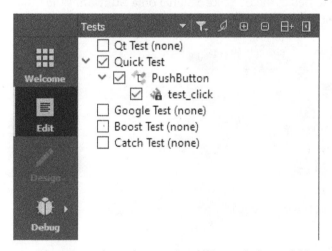

Figure 9.13 – The Tests explorer showing Quick Test with the available test cases

In this section, we discussed different testing approaches to test a QML object. In the next section, we will get familiar with GUI testing and learn about a few popular tools.

GUI testing tools

You can easily evaluate one or more classes as unit tests, but we have to manually write all of the test cases. GUI testing is an especially challenging task. How can we document user interactions such as mouse clicks without coding them in C++ or QML? This question has baffled developers. There are a number of GUI testing tools available on the market that help us do this. Some of them are expensive, some of them are open source. We will discuss a few such tools in this section.

However, you may not need a complete GUI testing framework. Some issues can be figured out with simple tricks. For example, while working with the GUI, you may also have to inspect different properties such as the alignment and boundaries of visual elements. One of the easiest ways is to add a `Rectangle` to inspect the boundary as shown in the next code:

```
Rectangle {
    id: container
    anchors {
        left: parent.left
        leftMargin: 100
        right: parent.right
        top: parent.top
        bottom: parent.bottom
    }
    Rectangle {
        anchors.fill : parent
        color: "transparent"
        border.color: "blue"        }
    Text {
        text: " Sample text"
        anchors.centerIn: parent
        Rectangle {
            anchors.fill : parent
            color: "transparent"
            border.color: "red"
        }
    }
}
```

When you run the preceding code snippet, you will see the GUI with element boundaries in colors as shown in the next screenshot:

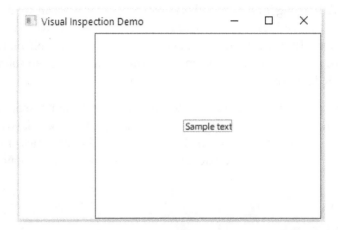

Figure 9.14 – Output of the visual boundaries of GUI elements using Rectangle

In the preceding example, you can see that the text element is placed centrally inside the rectangle with a blue border. Without the blue border, you might have wondered why it was not centrally placed in the GUI. You can also see the boundaries and margins of each element. When the text element width is less than the font width, then you will observe clipping. You can also find whether there are any overlapping regions between user interface elements. In this way, you can find issues in a specific element of the GUI without using the SG_VISUALIZE environment variable.

Let's discuss a few GUI testing tools.

The Linux Desktop Testing Project (LDTP)

The **Linux Desktop Testing Project (LDTP)** provides a high-quality test automation infrastructure and cutting-edge tools for testing and improving Linux desktop platforms. LDTP is a GUI testing framework that runs on all platforms. It pokes around in the application's user interface using the accessibility libraries. The framework also includes tools for recording test cases depending on how the user interacts with the GUI.

To click on a push button, use the following syntax:

```
click('<window name>','<button name>')
```

To get the current slider value of the given object, use the following code:

```
getslidervalue('<window name>','<slider name>')
```

To use LDTP for your GUI application, you must add an accessible name to all your QML objects. You can use object names as the accessible names as follows:

```
Button {
      id: quitButton
      objectName: "quitButton"
      Accessible.name: objectName
}
```

In the preceding code, we have added an accessible name to the QML control so that the LDTP tool can find this button. The LDTP requires the window name of the user interface to locate the child control. Let's say the window name is **Example**, then to generate a click event, use the following command on the LDTP script:

```
>click('Example','quitButton')
```

The preceding LDTP command locates the `quitButton` and generates a button-click event.

You can learn more about its features and uses at the following link:

```
https://ldtp.freedesktop.org/user-doc/
```

GammaRay

KDAB developed a software introspection tool named **GammaRay** to inspect Qt applications. You can observe and manipulate your application at runtime using the `QObject` introspection mechanism. This works on a local machine as well as a remote embedded target. It extends the capabilities of the instruction-level debugger while adhering to the same standards as the underlying frameworks. This is particularly useful for complex projects that use frameworks such as scene graphs, model/view, state machine, and so on. There are several tools available to inspect the objects and their properties. However, it stands out from other tools with its in-depth association with Qt's complex framework.

You can download GammaRay from the following link:

`https://github.com/KDAB/GammaRay/wiki/Getting-GammaRay`

You can learn more about its features and uses at the following link:

`https://www.kdab.com/development-resources/qt-tools/gammaray/`

Squish

Squish is a cross-platform GUI test automation tool for desktop, mobile, embedded, and web applications. You can automate GUI testing for your cross-platform application written with Qt Widgets or Qt Quick. Squish is used by thousands of organizations around the world to test their GUI with functional regression tests and system tests.

You can learn more about the tool at the following link:

`https://www.froglogic.com/squish/`

In this section, we discussed various GUI testing tools. Explore them and try them with your project. Let's summarize our learning in this chapter.

Summary

In this chapter, we have learned what debugging is and how to use different debugging techniques to identify technical issues in a Qt application. Apart from that, we've looked at the various debuggers that Qt supports on various operating systems. Finally, we learned how to use unit testing to simplify some of the debugging measures. We discussed unit testing, and you learned how to use the Qt Test framework. You saw how to debug a Qt Quick application. We also discussed various other testing frameworks and tools supported by Qt. Now you can write unit tests for your custom classes. The unit tests will fail and automatically alert if someone accidentally modifies some specific logic.

In *Chapter 10, Deploying Qt Applications*, you will learn about deploying Qt applications on various platforms. It will help you in creating installable packages for your target platform.

10
Deploying Qt Applications

In earlier chapters, you learned how to develop and test an application using Qt 6. Your application is ready and running on your desktop, but it is not standalone. You must follow certain steps to ship your application so that it can be used by end customers. This process is known as **deployment**. Generally, your end users want a single file that they can double-click to launch your software. Software deployment comprises different steps and activities that are required to make software available to its intended users who may not have any technical knowledge.

In this chapter, you will learn to deploy a Qt project on different platforms. Throughout, you will learn about the available deployment tools and important points to consider when creating a deployment package.

In this chapter, we will cover the following topics:

- Deployment strategies
- Static versus dynamic builds
- Deploying on desktop platforms
- Qt Installer framework
- Other installation tools
- Deploying on Android

By the end of this chapter, you will be able to create a deployable package and share it with others.

Technical requirements

The technical requirements for this chapter include minimum versions of Qt 6.0.0 and Qt Creator 4.14.0 installed on the latest desktop platform, such as Windows 10 or Ubuntu 20.04 or macOS 10.14.

All the code used in this chapter can be downloaded from the following GitHub link: `https://github.com/PacktPublishing/Cross-Platform-Development-with-Qt-6-and-Modern-Cpp/tree/master/Chapter10/HelloWorld`.

> **Important note**
> The screenshots used in this chapter are taken on the Windows platform. You will see similar screens based on the underlying platforms in your machine.

Understanding the need for deployment

The process of making software work on a target device, whether it's a test server, a production environment, a user's desktop, or mobile device, is known as **software deployment**. Typically, end users want a single file that they can open to access your application. The user will not want to go through several processes to obtain various alien files. Usually, users look for software that they can launch with a double click or tap. The user will not want to go through a series of steps to obtain a number of unknown files. In this chapter, we will discuss the steps and things to consider while deploying a Qt application. We will discuss deploying the application on Windows, Mac, Linux, and Android platforms.

We've just been running debug versions of the applications we've built so far. You should make release binaries for generating deployment packages. The difference between these two alternatives is that the debug version includes information about the code you write, making it much easier to debug if you encounter issues. However, you do not want to send multiple files to users because this is useless for them. Users just want to run your application. That is why you must provide them with your application's release version. So, to ship the app, we'll create it in release mode, which will give us a release binary that we can deliver to our users. Once you've got the binaries, you'll need to create separate packages depending on which platform you want to deploy your application. If you want to deploy on Windows, you're going to take a specific approach, and the same will apply to Linux, macOS, or Android.

A standard Qt deployment package consists of a single executable file, but it requires the presence of additional files in order to run. Aside from the executable file, the following files will be required:

- Dynamic libraries
- Third-party libraries
- Add-on modules
- Distributable files
- Qt plugins
- Translation files
- Help files
- Licenses

When we start a Qt project in Qt Creator, it is set to use dynamic linking by default. Therefore, our app will require the Qt dynamic link libraries. We will also require C++ runtime of your favorite compiler (MinGW/MSVC/Clang/GCC) and standard library implementations. These are usually available as `.dll` file on Windows, `.so` file on Linux and `.so`, or `.dylib` file on macOS. If your project is a large complex project, you may have multiple libraries. Your application package may also require third-party libraries such as opengl, libstdc++, libwinpthread, and openssl.

If your application is based on Qt Quick, then you will also require standard modules such as QtQuick, QtQml, QtStateMachine, QtCharts, and Qt3D. They are supplied as dynamic libraries, with some extra files providing QML module metadata, or as pure QML files. Unfortunately, the dynamic libraries that implement Qt's C++ and QML APIs are insufficient to allow our executable to run. Qt also uses plugins to enable extensions, as well as for fairly standard GUI capabilities such as image file loading and display. Similarly, some plugins encapsulate the platform on which Qt runs.

If you are using Qt's translation support, then you will also require the translation files to be deployed. We will discuss translation more in *Chapter 11, Internationalization*. You may also need to deploy the documentation files if you are using the Qt Help framework or even simple PDF manuals. You may also need to deploy some icons or script or license agreements for your application. You also have to ensure that the Qt libraries can locate the platform plugins, documentation, and translations, as well as the intended executable file, by themselves.

Choosing between static and dynamic libraries

You can build your Qt application using static linking or dynamic linking. When you build an application, the linker adds the information about dependent libraries or functions to the executable using either of these two approaches. We assume that you are already aware of these two methods. In this section, we will discuss when to use static linking and when to use dynamic linking for your Qt application.

Static Library, or **statically linked library**, originates from the linker putting all required library functions to the executable file. Static linking generates bigger binary files that require more storage and main memory space. Static libraries are represented by the `.a` file extension in Linux and the `.lib` file extension in Windows.

Dynamic Library, or **dynamically linked shared library**, does not need the code to be transferred. Instead, the name of the library is simply included in the binary file. When an application is launched, both the binary file and the library are loaded into memory. Dynamic libraries are linked at runtime. They are represented by the `.so` file extension in Linux and the `.dll` file extension in Windows.

A static build consists of a single executable file. But in a dynamic build, you must take care of the dynamic libraries. Static builds are simpler as they may already have the Qt plugins and QML imports in the executable. The static build also facilitates **link time optimization** (**LTO**), which can improve the overall application performance. Since it avoids the burden of downloading the Qt libraries and ensuring that they are located in the default search path for libraries on the target system, static linking is frequently the safest and easiest approach to publish an application. However, static linking is not very useful unless Qt was built from a source with the `-static` configuration option specified. This mode of Qt application deployment is available only with a commercial license. You should avoid linking your application statically if you are an open source developer. Since we are using an open source Qt version in this book, we won't go through static builds in any more detail. Instead, we'll stick to the regular dynamic builds and deployments.

You can learn more about deploying a Qt application with the aforementioned approaches at the following link:

`https://doc.qt.io/qt-6/deployment.html`.

In the following sections, we are going to be focusing on the leading desktop and mobile platforms. We're not going to discuss embedded platforms as this is beyond the scope of this book.

Deploying on desktop platforms

You have seen that there's a lot to consider when deploying a Qt application. Fortunately, Qt provides a tool that can assist us in this process by scanning the generated application binary, identifying all dependencies, and copying them to the deployment directory. We will deploy our application on various platforms to achieve different objectives, but the concepts will remain the same. Once we have our binary built, the first thing we need to do is add the dependencies so that the user can execute the application without difficulties.

There are two ways in which we may load the dependencies. We can do it manually or use certain tools that are provided by the Qt framework or by a third party. On Windows, we can use `windeployqt` to load our dependencies. On macOS, we can use `macdeployqt` to load our dependencies for our binary. There is also another tool called `linuxdeployqt` that you can use to add the dependencies to your binary. `linuxdeployqt` works well for our needs, and we're going to discuss it in this chapter. However, this Linux deploy utility tool is not official and is not supported by Qt. Once you have your binary generated, you need to find and add in the dependencies. You can do that manually or use one of these tools, depending on where you are, to deploy your application.

In this chapter, we will use a simple *HelloWorld* example to discuss how to deploy applications on different platforms. We will find the dependencies and create a standalone package. Let's begin with Windows deployment.

Deploying on Windows

Most of the desktop applications that are built for Windows are usually delivered in two approaches. First, the application comes as a standalone application without any need for installation. In this approach, the application usually come as an executable file (`.exe`) with all dependent libraries inside the same directory. This type of application is known as a **portable application**. The application doesn't make an entry into the installed application list. So, you won't find an option in the add or remove program list. This is useful when you don't have permission to install a new application on the Windows desktop. The second type of application usually comes in `.exe` or `.msi` format. You will learn to create an installable `.exe` file. In this section, we will discuss how to create standalone deployment packages with both approaches.

Follow these steps to create a portable application:

1. First, create a simple Qt application. You can choose Qt Widget or a Qt Quick-based application. Here we will discuss the Qt Widget-based application. The procedure is the same for both types of applications.

2. Once you created the sample application, you can optionally add your application name, version, organization name, and domain by adding the few lines of code inside your `main.cpp` file, as shown here:

```
QApplication app (argc, argv);
app.setOrganizationName("Awesome Company");
app.setOrganizationDomain("www.abc.com");
app.setApplicationName("Deployment Demo");
app.setApplicationVersion("1.0.0");
```

3. Once you created the application, build it in **Release** mode. You can change the **Build** mode in the build settings. **Release** mode creates a smaller binary as it eliminates the debug symbols. You can quickly change the build mode from the kit selector section by clicking on it and then selecting the **Release** option, as shown in *Figure 10.1*:

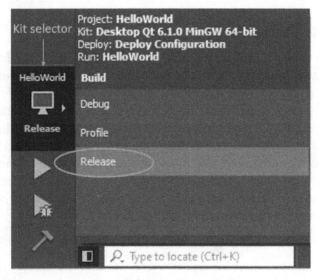

Figure 10.1 – Release option in Qt Creator

4. You can see that the binaries are created inside the **Release** directory. In this example, we are using *shadow build*. You can also change the release directory from the **General** section under the **Build Settings** screen:

Name	Date modified	Type	Size
HelloWorld.exe	18-05-2021 22:59	Application	24 KB
main.o	18-05-2021 22:59	O File	2 KB
moc_predefs.h	18-05-2021 22:59	C++ Header file	16 KB
moc_widget.cpp	18-05-2021 22:59	C++ Source file	3 KB
moc_widget.o	18-05-2021 22:59	O File	9 KB
widget.o	18-05-2021 22:59	O File	4 KB

« build-HelloWorld-Desktop_Qt_6_1_0_MinGW_64_bit-Release › release

Figure 10.2 – Directory with release binaries

5. Now, create a deployment directory and copy the executable from the **Release** directory.

6. Now, double-click on the executable file. You will notice that the application failed to launch and that several error dialogs appear. The error dialogs will mention which library is missing. If you don't see these errors, then you might have already added the library path in the system environment. You can try it on a clean system where Qt libraries are not installed:

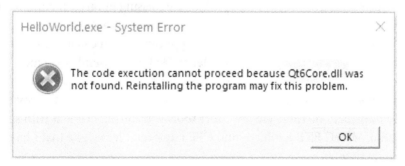

HelloWorld.exe - System Error   ✕

The code execution cannot proceed because Qt6Core.dll was not found. Reinstalling the program may fix this problem.

OK

Figure 10.3 – Error showing Qt library dependency

7. The next step is to find the missing Qt libraries that are required to run your application independently outside the IDE.

8. Since we are using an open source version of Qt and the dynamic linking approach here, you will notice that the missing libraries will have a `.dll` extension. Here, we saw that the missing library is `Qt6Core.dll`.

9. The number of errors will depend on the number of modules used
    in the program. You can find the Qt dependent libraries from the
    `QTDIR/6.x.x/<CompilerName>/bin` directory. Here, `QTDIR` is where Qt 6 is
    installed. In our example, we have used *Qt 6.1.0* as the version and *mingw81_64* as
    the compiler, hence the path is `D:/Qt/6.1.0/mingw81_64/bin`. This path can
    vary as per your Qt installation path, Qt version, and chosen compiler. The following
    screenshot shows the presence of the dynamic libraries under the `bin` directory:

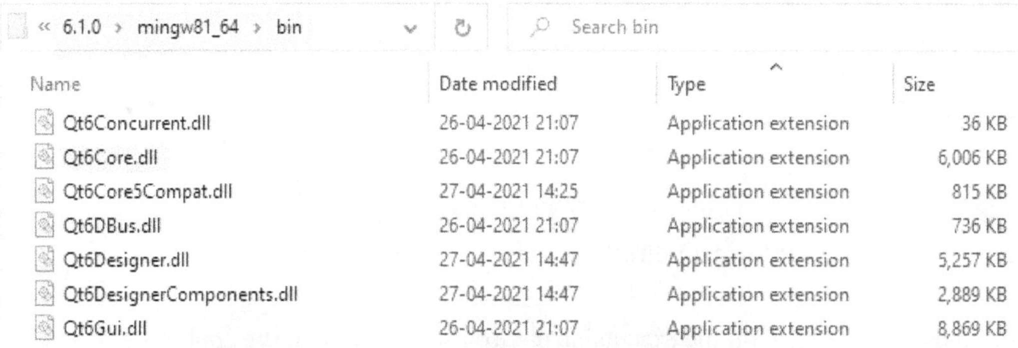

Name	Date modified	Type	Size
Qt6Concurrent.dll	26-04-2021 21:07	Application extension	36 KB
Qt6Core.dll	26-04-2021 21:07	Application extension	6,006 KB
Qt6Core5Compat.dll	27-04-2021 14:25	Application extension	815 KB
Qt6DBus.dll	26-04-2021 21:07	Application extension	736 KB
Qt6Designer.dll	27-04-2021 14:47	Application extension	5,257 KB
Qt6DesignerComponents.dll	27-04-2021 14:47	Application extension	2,889 KB
Qt6Gui.dll	26-04-2021 21:07	Application extension	8,869 KB

Figure 10.4 – Required Qt libraries inside the bin directory

10. As illustrated in *Figure 10.4*, copy the missing `.dll` files to the recently created
    deployment directory.

11. Repeat the process until you have copied all the missing libraries mentioned in
    the error messages to the deployment directory. You may also have to deploy
    compiler-specific libraries along with your application. You can also find the
    dependent libraries by using the **Dependency Walker (depends.exe)** tool. This tool
    is a free tool specific to Windows. It provides a list of dependent libraries. However,
    in recent versions, the tool has not been very useful and often fails to provide the
    required information. There are few more tools you can experiment with such
    as PeStudio, MiTeC EXE Explorer, and CFF Explorer. Please note that I haven't
    explored these tools.

12. Once you have copied all the missing libraries, try to run the application again. This time, you will notice that a new error pops up. On this occasion, the message relates to the platform plugin:

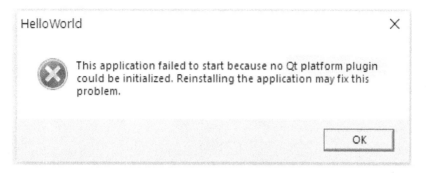

Figure 10.5 – Error dialog indicating a missing Qt platform plugin

13. Create a directory called `platforms` inside the deployment directory:

Name	Date modified	Type	Size
qwindows.dll	26-04-2021 21:08	Application extension	975 KB
qoffscreen.dll	26-04-2021 21:08	Application extension	83 KB
qminimal.dll	26-04-2021 21:08	Application extension	42 KB
qdirect2d.dll	26-04-2021 21:08	Application extension	1,064 KB
qwindows.debug	26-04-2021 21:00	DEBUG File	68,848 KB
qdirect2d.debug	26-04-2021 20:58	DEBUG File	78,123 KB
qoffscreen.debug	26-04-2021 20:58	DEBUG File	5,090 KB
qminimal.debug	26-04-2021 20:58	DEBUG File	3,533 KB

Figure 10.6 – Directory showing the Qt windows platform plugin

14. Then, copy the `qwindows.dll` file from `C:\Qt\6.x.x\<compiler_name>\plugins\platforms` to the new `platforms` subdirectory. *Figure 10.7* illustrates the organization of the files in the deployment directory:

Name	Date modified	Type	Size
« release › deployment		∨ ↻	🔎 Search deployment
platforms	18-05-2021 22:53	File folder	
HelloWorld.exe	18-05-2021 22:39	Application	24 KB
libgcc_s_seh-1.dll	12-05-2018 11:41	Application exten...	75 KB
libstdc++-6.dll	12-05-2018 11:41	Application exten...	1,384 KB
libwinpthread-1.dll	12-05-2018 11:41	Application exten...	51 KB
Qt6Core.dll	26-04-2021 21:07	Application exten...	6,006 KB
Qt6Gui.dll	26-04-2021 21:07	Application exten...	8,869 KB
Qt6Widgets.dll	26-04-2021 21:07	Application exten...	6,219 KB

Figure 10.7 – Copy platforms plugin inside the release directory

15. Now, double-click on the `HelloWorld.exe` file. You will observe that the **HelloWorld!** GUI appears in no time. Now, the Qt Widgets application can be launched on a Windows platform that doesn't have Qt 6 installed:

Figure 10.8 – Standalone application running with resolved dependencies

16. The next and final step is to zip the folder and share it with your friends.

Congratulations! You have successfully deployed your first standalone application. However, this approach doesn't work well for a large project where we have many dependent files. Qt provides several handy tools for dealing with such challenges and creating an installation package easily. In the next section, we will discuss the Windows deployment tool and how it can help us deal with these challenges.

Windows deployment tool

The Windows deployment tool comes with the Qt 6.x installation package. You can find it under <QTDIR>/bin/ as windeployqt.exe. You can run this tool from the Qt command prompt and pass the executable file as the argument, or with a directory as the parameter. If you are building a Qt Quick application, you will have to additionally add the directory path of the .qml files.

Let's have a look at some of the important command-line options available in windeployqt. Explore some of the useful options in the following list:

- -? or -h or --help displays help on command-line options.
- --help-all displays help including Qt-specific options.
- --libdir <path> copies dependent libraries to the path.
- --plugindir <path> copies dependent plugins to the path.
- --no-patchqt instructs not to patch the Qt6Core library.
- --no-plugins instructs to skip plugin deployment.
- --no-libraries instructs to skip library deployment.
- --qmldir <directory> scans the QML imports from the source directory.
- --qmlimport <directory> adds the given path to the QML module search locations.
- --no-quick-import instructs to skip deployment of Qt Quick imports.
- --no-system-d3d-compiler instructs to skip deployment of the D3D compiler.
- --compiler-runtime deploys the compiler runtime on the desktop.

- `--no-compiler-runtime` prevents deployment of the compiler runtime on the desktop.

- `--no-opengl-sw` prevents deployment of the software rasterizer library.

You can find the `windeployqt` tool inside the `bin` folder, as shown in the following screenshot:

Name	Date modified	Type	Size
windeployqt.exe	27-04-2021 14:49	Application	225 KB
uic.exe	26-04-2021 21:08	Application	570 KB
tracegen.exe	26-04-2021 21:08	Application	1,578 KB
shadergen.exe	27-04-2021 14:52	Application	262 KB
rcc.exe	26-04-2021 21:08	Application	1,795 KB
qvkgen.exe	26-04-2021 21:08	Application	77 KB
qtplugininfo.exe	27-04-2021 14:49	Application	41 KB
qtpaths.exe	27-04-2021 14:48	Application	70 KB

Search bin — « Qt > 6.1.0 > mingw81_64 > bin

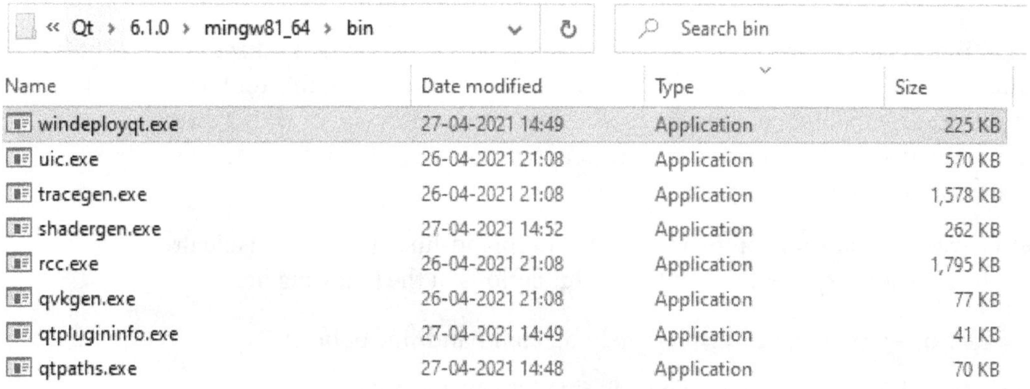

Figure 10.9 – The windeployqt tool inside the bin directory

The easiest way to use `windeployqt` is to add its path to the **Path** variable. To add it to **Path**, open **System Properties** on your Windows machine and then click on **Advance system settings**. You will find that the **System Properties** window appears. At the bottom of the **System Properties** window, you will see the **Environment Variables…** button. Click it and select the **Path** variable, as shown in the following screenshot. Then, click on the **Edit…** button. Add the path of the Qt bin directory and hit the **OK** button:

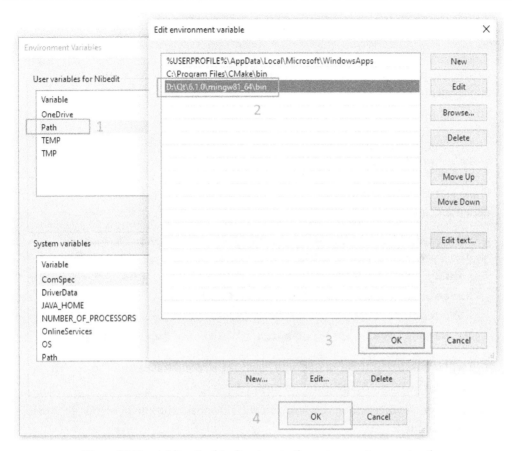

Figure 10.10 – Adding the bin directory to the system environment path

Close the **System Properties** screen and launch the Qt command prompt. Then you can use the following syntax to create a deployment package for your Qt Widget-based application:

```
>windeployqt <your-executable-path>
```

If you are using Qt Quick, follow the next syntax:

```
>windeployqt --qmldir <qmlfiles-path> <your-executable-path>
```

Afterward, the tool will copy the identified dependencies to the deployment directory, ensuring that we have all of the required components in one location. It will also build the subdirectory structure for plugins and other Qt resources that you'd expect. If ICU and other files are not in the bin directory, they must be added to the **Path** variable before running the tool.

Let's begin with the same *HelloWorld* example. To create a deployment of the example using `windeployqt`, perform the following steps:

1. Create a deployment directory and copy the `HelloWorld.exe` file to the deployment directory.

2. Now you can invoke the deployment tool, as shown here:

   ```
   D:\Chapter10\HelloWorld\deployment>windeployqt
   HelloWorld.exe
   ```

3. Once you enter the command, the tool will start gathering information about the dependencies:

   ```
   >D:\Chapter10\HelloWorld\deployment\HelloWorld.exe 64
   bit, release executable
   Adding Qt6Svg for qsvgicon.dll
   Direct dependencies: Qt6Core Qt6Widgets
   All dependencies   : Qt6Core Qt6Gui Qt6Widgets
   To be deployed     : Qt6Core Qt6Gui Qt6Svg Qt6Widgets
   ```

4. You will observe that the tool not only listed the dependencies but also copied the required files to the destination directory.

5. Open the deployment directory and you will find that multiple files and directories have been added:

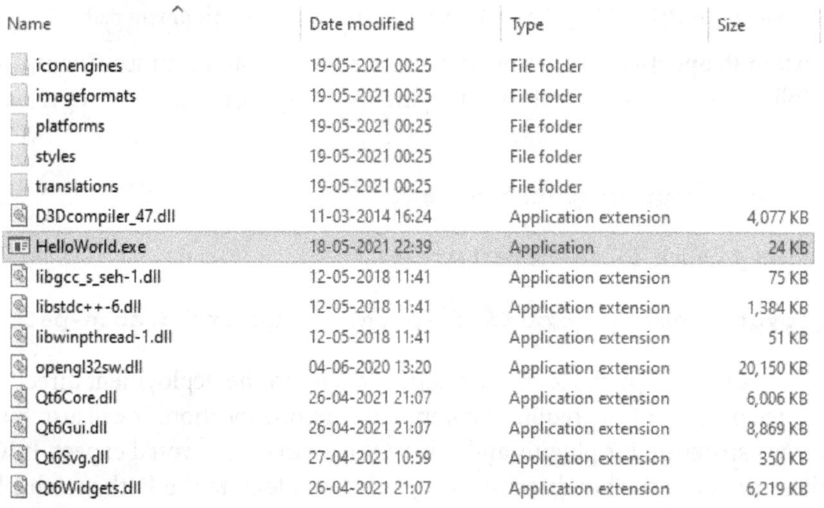

Name	Date modified	Type	Size
iconengines	19-05-2021 00:25	File folder	
imageformats	19-05-2021 00:25	File folder	
platforms	19-05-2021 00:25	File folder	
styles	19-05-2021 00:25	File folder	
translations	19-05-2021 00:25	File folder	
D3Dcompiler_47.dll	11-03-2014 16:24	Application extension	4,077 KB
HelloWorld.exe	18-05-2021 22:39	Application	24 KB
libgcc_s_seh-1.dll	12-05-2018 11:41	Application extension	75 KB
libstdc++-6.dll	12-05-2018 11:41	Application extension	1,384 KB
libwinpthread-1.dll	12-05-2018 11:41	Application extension	51 KB
opengl32sw.dll	04-06-2020 13:20	Application extension	20,150 KB
Qt6Core.dll	26-04-2021 21:07	Application extension	6,006 KB
Qt6Gui.dll	26-04-2021 21:07	Application extension	8,869 KB
Qt6Svg.dll	27-04-2021 10:59	Application extension	350 KB
Qt6Widgets.dll	26-04-2021 21:07	Application extension	6,219 KB

Figure 10.11 – windeployqt copied all required files to the deployment directory

6. In the previous section, we had to identify and copy all the dependencies ourselves, but that task is now delegated to the `windeployqt` tool.

7. If you are using a *Qt Quick application*, run the following command:

```
>D:\Chapter10\qmldeployment>windeployqt.exe --qmldir D:\
Chapter10\HelloWorld D:\Chapter10\qmldeployment
```

8. You can see that the tool has gathered the dependencies and copied the required files to the deployment directory:

```
D:\Chapter10\qmldeployment\HelloWorld.exe 64 bit, release
executable [QML]

Scanning D:\Chapter10\HelloWorld:

QML imports:
   'QtQuick' D:\Qt\6.1.0\mingw81_64\qml\QtQuick

   'QtQuick.Window' D:\Qt\6.1.0\mingw81_64\qml\QtQuick\
Window

   'QtQml' D:\Qt\6.1.0\mingw81_64\qml\QtQml

   'QtQml.Models' D:\Qt\6.1.0\mingw81_64\qml\QtQml\Models

   'QtQml.WorkerScript' D:\Qt\6.1.0\mingw81_64\qml\QtQml\
WorkerScript

Adding Qt6Svg for qsvgicon.dll

Direct dependencies: Qt6Core Qt6Gui Qt6Qml

All dependencies   : Qt6Core Qt6Gui Qt6Network Qt6OpenGL
Qt6Qml Qt6Quick Qt6QuickParticles Qt6Sql

To be deployed     : Qt6Core Qt6Gui Qt6Network Qt6OpenGL
Qt6Qml Qt6Quick Qt6QuickParticles Qt6Sql Qt6Svg
```

9. You can now double-click to launch your standalone application.

10. The next step is to zip the folder and share it with your friends.

The command-line options for the Windows deployment tool can be used to fine-tune the identification and copy process. The essential instructions may be found at the following links:

https://doc.qt.io/qt-6/windows-deployment.html.

https://wiki.qt.io/Deploy_an_Application_on_Windows.

Cheers! You have learned to deploy a Qt application using the Windows deployment tool. However, there is still a lot of work to be done. The Qt Installer Framework provides several handy tools for dealing with such challenges and creating installable packages easily. In the next section, we will discuss the Linux deployment tool and how to use it to create a standalone application.

Deploying on Linux

In Linux distributions, we have multiple options to deploy our application. You can use an installer, but you can also have an option such as an app bundle. There is a technology called **app image** that makes the deployment process very easy. You can also package your application to be used by the package manager. You can go through options such as apt on Debian, Ubuntu, or Fedora and your application can be used through this approach. However, you can choose a much simpler approach, like the app image option, which is going to provide you with one file. You can give that file to your users and they can just double-click to run the application.

Qt documentation provides certain instructions to deploy on Linux. You can have a look at the following link:

https://doc.qt.io/qt-6/linux-deployment.html.

Qt doesn't provide any ready-made tool similar to windeployqt for Linux distributions. This may be due to a large number of Linux flavors. However, there is an unofficial open source Linux deployment tool named linuxdeployqt. This takes an application as input and turns it into a self-contained package by replicating the project's resources into a bundle. Users can get the generated bundle as an AppDir or an AppImage, or it may be included in cross-distribution packages. With systems such as CMake, qmake, and make, it may be used as part of the build process to deploy applications written in C, C++, and other compiled languages. It can package specific libraries and components required to run the Qt-based application.

You can download linuxdeployqt from the following link:

https://github.com/probonopd/linuxdeployqt/releases.

You will get linuxdeployqt-x86_64.AppImage after the download and do chmod a+x before running it.

You can read the complete documentation and find the source code at `https://github.com/probonopd/linuxdeployqt`.

If you want a single application package easily, then run `linuxdeployqt` with the `-appimage` flag.

There are also few more deployment tools such as **Snap** and **Flatpak** to package an application and its dependencies runs across multiple Linux distributions without making any modification.

You can read on how to create a snap in the following link: `https://snapcraft.io/docs/creating-a-snap`

You can explore more about Flatpak by visiting the following link: `https://docs.flatpak.org/en/latest/qt.html`

In the next section, we will discuss the macOS deployment tool and how to use it to create a standalone application for your Mac users.

Deploying on macOS

You can go through a similar process as discussed in previous sections to generate an installer file for the macOS. We will discuss the steps that you can follow to generate an app bundle. You can test the package on your macOS and send it to your Mac users. The process is pretty much the same as on Linux. After all, macOS is based on Unix. Therefore, you can create installers, which we call a bundle on the macOS.

You can find the macOS deployment tool inside `QTDIR/bin/macdeployqt`. It is designed to automate the process of creating a deployable application bundle that contains the Qt libraries as private frameworks. The Mac deployment tool also deploys the Qt plugins unless you specify the `-no-plugins` option. By default, Qt plugins such as platform, image format, print support, and accessibility are always deployed. SQL driver and SVG plugins are deployed only if it is used by the application. The designer plugins are not deployed. If you want to include a third-party library in the application bundle, you must manually copy the library into the bundle after it has been built.

A couple of years back, Apple launched a new filesystem called **Apple File System (APFS)**. Older versions of macOS cannot read APFS-formatted `.dmg` files. For compatibility with all versions of macOS currently supported by Qt, `macdeployqt` uses the older HFS+ filesystem by default. To choose a different filesystem, use the `-fs` option.

You can find detailed instructions at the following link: `https://doc.qt.io/qt-6/macos-deployment.html`.

In the next section, we will discuss the Qt Installer Framework and how to use it to create a complete installation package for your users.

Using the Qt Installer Framework

The **Qt Installer Framework** (**QIFW**) is a collection of cross-platform tools and utilities for creating installers for the supported desktop Qt platforms, which include Linux, Windows, and macOS. It allows you to distribute your application across all supported desktop Qt platforms without having to rewrite the source code. The Qt Installer Framework tools create installers that include a collection of pages that help users through the installation, update, and removal processes. You provide the installable contents as well as information about it, such as the product name, the installer, and the legal agreement.

You may personalize the installers by adding widgets to the pre-defined pages or adding entire pages to give consumers more options. You may add operations to the installer by writing scripts. Depending on your use cases, you can give end customers an offline or online installation, or both. It works well on Windows, Linux, and Mac. We will use it to create installers for our application and we're going to see how this works in detail on Windows. The process followed for Linux and macOS is similar to Windows. So, we will only discuss the Windows platform. You can try similar steps on your favorite platform.

You can learn more about the predefined pages at the following link: `https://doc.qt.io/qtinstallerframework/ifw-use-cases-install.html`.

Before starting the journey, confirm that Qt Installer Framework is installed on your machine. If it is not present, launch **Qt Maintenance Tool** and install it from the **Select Components** page, as shown in the following screenshot:

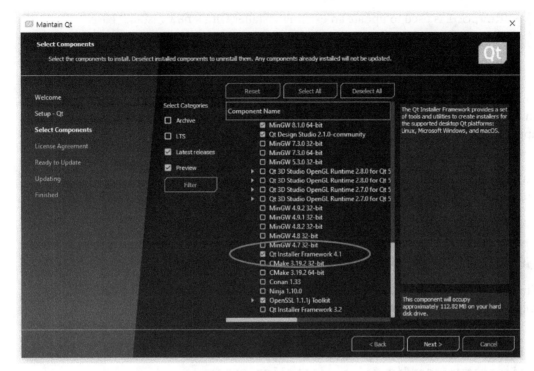

Figure 10.12 – The Qt Installer Framework download option in the Qt maintenance tool

Once you have installed the application successfully, you will find the installation files under QTDIR\Tools\QtInstallerFramework\:

Name	Date modified	Type	Size
archivegen.exe	10-03-2020 15:28	Application	19,694 KB
binarycreator.exe	10-03-2020 15:28	Application	20,303 KB
devtool.exe	10-03-2020 15:29	Application	20,151 KB
installerbase.exe	10-03-2020 15:28	Application	21,034 KB
repogen.exe	10-03-2020 15:28	Application	20,078 KB

Figure 10.13 – Tools inside the Qt Installer Framework directory on Windows

You can see that there are five executables created inside the Qt Installer Framework directory:

- The archivegen tool is used to package files and directories into 7zip archives.

- The binarycreator tool is used to create online and offline installers.

- devtool is used to update an existing installer with a new installer base.

- The installerbase tool is the core installer that packs all data and meta information.

- The repogen tool is used to generate online repositories.

In this section, we will use the binarycreator tool to create the installer for our Qt application. This tool may be used to produce both offline and online installers. Some choices have default values, so you may leave them out.

To create an offline installer on your Windows machine, you can enter the following command to your Qt command prompt:

```
><location-of-ifw>\binarycreator.exe -t <location-of-
ifw>\installerbase.exe -p <package_directory> -c <config_
directory>\<config_file> <installer_name>
```

Similarly, to create an offline installer on your Linux or Mac machine, you can enter the following command to your Qt command prompt:

```
><location-of-ifw>/binarycreator -t <location-of-
ifw>/installerbase -p <package_directory> -c <config_
directory>/<config_file> <installer_name>
```

For example, to create an offline installer, execute the following command:

```
>binarycreator.exe --offline-only -c installer-config\config.
xml -p packages-directory -t installerbase.exe SDKInstaller.exe
```

The preceding instruction will create an offline installer for the SDK, containing all dependencies.

To create an online-only installer, you may use --online-only, which defines all the packages to install from an online repository on a web server. For example, to create an online installer, execute the following command:

```
>binarycreator.exe -c installer-config\config.xml -p packages-
directory -e org.qt-project.sdk.qt,org.qt-project.qtcreator -t
installerbase.exe SDKInstaller.exe
```

You can learn more about binarycreator and the different options at the following page: https://doc.qt.io/qtinstallerframework/ifw-tools.html#binarycreator.

The easiest way to use binarycreator is to add its path to the **Path** variable. To add it to **Path**, open **System Properties** on your Windows machine and then click on **Advance system settings**. You will find that the **System Properties** window appears. At the bottom of the **System Properties** window, you will see the **Environment Variables…** button. Click it and select the **Path** variable, as shown in the following screenshot. Then, click on the **Edit…** button. Add the path of the QIFW bin directory and hit the **OK** button. The following screenshot illustrates how to do this:

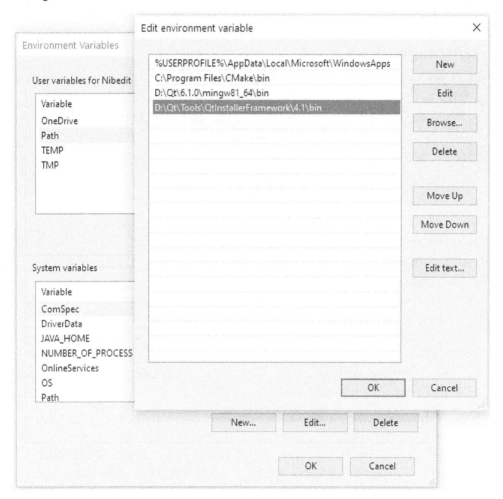

Figure 10.14 – Adding the QIFW bin directory to the system environment path

Close the **System Properties** screen and launch the Qt command prompt.

Let's proceed to deploy our sample *Hello World* application. We're going to create one installable package for our users so that they can double-click and install it:

1. Create a directory structure that matches the installer's design and allows it to be extended in the future. The `config` and `packages` subdirectories must be present in the directory. It doesn't matter where you put the directory for QIFW deployment; all that matters is that it has this structure.

2. Make a configuration file with instructions for building the installer binaries and online repositories. Create a file called `config.xml` in the config directory with the following content:

```xml
<?xml version="1.0" encoding="UTF-8"?>
<Installer>
    <Name>Deployment Example </Name>
    <Version>1.0.0</Version>
    <Title>Deployment Example</Title>
    <Publisher>Packt</Publisher>
    <StartMenuDir>Qt6 HelloWorld</StartMenuDir>
    <TargetDir>@HomeDir@/HelloWorld</TargetDir>
</Installer>
```

The `Title` tag gives the name of the installer that appears in the title bar. The application name is added to the page name and introductory text using the `Name` tag. The software version number is specified by the `Version` tag. The `Publisher` tag defines the software's publisher. The name of the default program group for the product in the Windows Start menu is specified by the `StartMenuDir` tag. The default destination directory presented to users is `InstallationDirectory` in the current user's home directory, as specified by the `TargetDir` tag. You can read about more tags in the documentation.

You can also specify the app bundle icon in `config.xml`. On Windows, it is extended with `.ico` and can be used as the application icon for the `.exe` file. On Linux, you can specify the icon with a `.png` extension and this can be used as the window icon. On macOS, you can specify the icon with `.icns` and this can be used as the icon for the newly produced bundle.

3. Now create a subdirectory inside the `packages` directory. This will be your `component name`. You can use your organization's name and application name or your organization domain as the `component` such as `CompanyName.ApplicationName`. The directory name functions as a domain-like identifier, identifying all components.

4. Make a package information file with details about the components that may be installed. In this simple example, the installer just has to deal with one component. Let's create a package information file called `package.xml` inside the `packages\{component}\meta` directory.

5. Add the file in side the meta-directory with the following content to provide information about the component to the installater.

   Let's create `package.xml` and add the following content to it:

```xml
<?xml version="1.0"?>
<Package>
    <DisplayName>Hello World</DisplayName>
    <Description>This is a simple deployment example.
    </Description>
    <Version>1.0.1</Version>
    <ReleaseDate>2021-05-19</ReleaseDate>
</Package>
```

   The information from the following elements is displayed on the component selection page during installation:

- The `DisplayName` tag specifies the name of the component in the list of components.

- The `Description` tag specifies the text that is displayed when the component is selected.

- The `Version` tag enables you to promote updates to users when they become available.

- The `Default` tag specifies whether the component is selected by default. The value `true` sets the component as selected.

- You can add licensing information to your installer. The name of the file that includes the text for the licensing agreement that is shown on the licensing check page is specified by the `License` tag.

6.   You can copy the required content inside the `data` subdirectory under the `package` directory. Copy all the files and directories to the `data` subdirectory, which were earlier created with `windeployqt`. The following screenshot shows the content copied inside the `data` subdirectory:

Name	Date modified	Type	Size
iconengines	19-05-2021 00:25	File folder	
imageformats	19-05-2021 00:25	File folder	
platforms	19-05-2021 00:25	File folder	
styles	19-05-2021 00:25	File folder	
translations	19-05-2021 00:25	File folder	
D3Dcompiler_47.dll	11-03-2014 16:24	Application extension	4,077 KB
HelloWorld.exe	18-05-2021 22:39	Application	24 KB
libgcc_s_seh-1.dll	12-05-2018 11:41	Application extension	75 KB
libstdc++-6.dll	12-05-2018 11:41	Application extension	1,384 KB
libwinpthread-1.dll	12-05-2018 11:41	Application extension	51 KB
opengl32sw.dll	04-06-2020 13:20	Application extension	20,150 KB
Qt6Core.dll	26-04-2021 21:07	Application extension	6,006 KB
Qt6Gui.dll	26-04-2021 21:07	Application extension	8,869 KB
Qt6Svg.dll	27-04-2021 10:59	Application extension	350 KB
Qt6Widgets.dll	26-04-2021 21:07	Application extension	6,219 KB

Figure 10.15 – Contents generated by windeployqt copied inside the data subdirectory

7.   The next step is to use the `binarycreator` tool to create the installer. Enter the following instruction to the Qt command prompt:

```
>binarycreator.exe -c config/config.xml -p packages
HelloWorld.exe
```

8.   You can see that an installer file got generated inside our deployment directory:

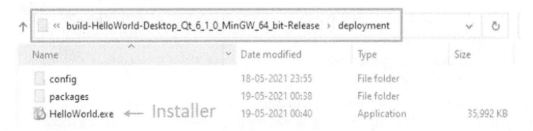

Figure 10.16 – Installer package created inside the deployment directory

You can also follow the same steps and run the following command to generate an installer file on Linux or macOS:

```
$./binarycreator -c config/config.xml -p packages
HelloWorld
```

9. We have got the desired result. Now, let's run the installer to verify that the deployment package has been created properly.

10. Double-click on the installer file to begin the installation. You will see a nice setup wizard appear on screen:

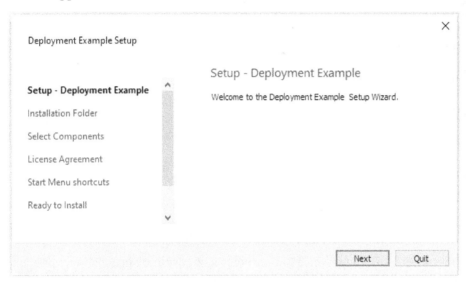

Figure 10.17 – Setup wizard running a deployment example

11. Proceed through the pages to complete the installation. Exit the setup wizard.

12. Now, launch the application from the Windows **Start** menu. You should see the **HelloWorld** user interface appear in no time.

13. You can also find the installed application in **Add/Remove Programs**:

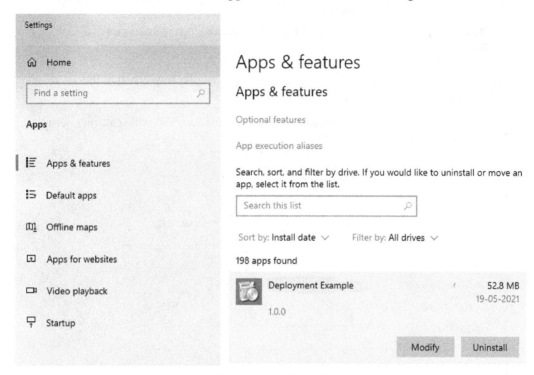

Figure 10.18 – Deployment example entry in the Windows program list

14. You may use the maintenance tool installed along with the installation package to update, uninstall, and add components to the application. You can find the tool inside your installation directory, as shown in the following screenshot:

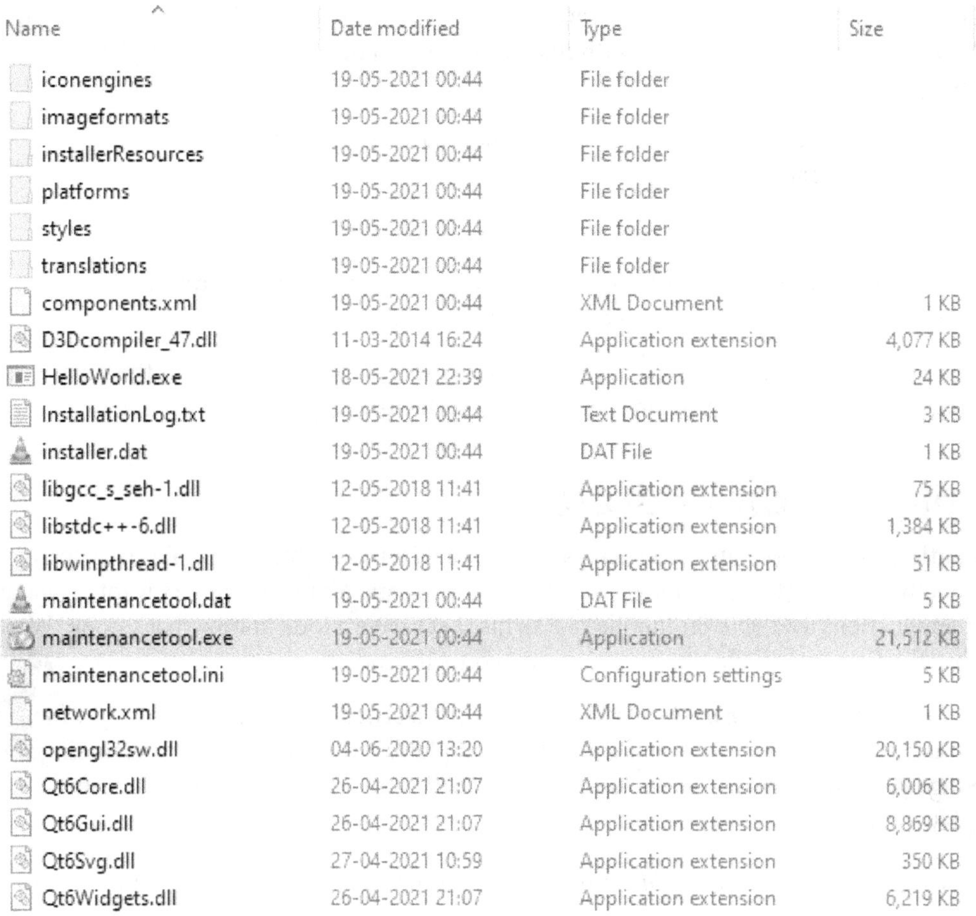

Name	Date modified	Type	Size
iconengines	19-05-2021 00:44	File folder	
imageformats	19-05-2021 00:44	File folder	
installerResources	19-05-2021 00:44	File folder	
platforms	19-05-2021 00:44	File folder	
styles	19-05-2021 00:44	File folder	
translations	19-05-2021 00:44	File folder	
components.xml	19-05-2021 00:44	XML Document	1 KB
D3Dcompiler_47.dll	11-03-2014 16:24	Application extension	4,077 KB
HelloWorld.exe	18-05-2021 22:39	Application	24 KB
InstallationLog.txt	19-05-2021 00:44	Text Document	3 KB
installer.dat	19-05-2021 00:44	DAT File	1 KB
libgcc_s_seh-1.dll	12-05-2018 11:41	Application extension	75 KB
libstdc++-6.dll	12-05-2018 11:41	Application extension	1,384 KB
libwinpthread-1.dll	12-05-2018 11:41	Application extension	51 KB
maintenancetool.dat	19-05-2021 00:44	DAT File	5 KB
maintenancetool.exe	19-05-2021 00:44	Application	21,512 KB
maintenancetool.ini	19-05-2021 00:44	Configuration settings	5 KB
network.xml	19-05-2021 00:44	XML Document	1 KB
opengl32sw.dll	04-06-2020 13:20	Application extension	20,150 KB
Qt6Core.dll	26-04-2021 21:07	Application extension	6,006 KB
Qt6Gui.dll	26-04-2021 21:07	Application extension	8,869 KB
Qt6Svg.dll	27-04-2021 10:59	Application extension	350 KB
Qt6Widgets.dll	26-04-2021 21:07	Application extension	6,219 KB

Figure 10.19 – Maintenance tool in the installation directory

Congratulations! You have created an installer package for your sample application. Now you can ship your developed Qt application to your users and friends.

You can also provide further customization with customized setup wizard pages. You can find the complete list of templates with installers that can be used with the QIFW at the following link:

`https://doc.qt.io/qtinstallerframework/ifw-customizing-installers.html`

`https://doc.qt.io/qtinstallerframework/qtifwexamples.html.`

You can explore more features of the framework here: `https://doc.qt.io/qtinstallerframework/ifw-overview.html.`

In this section, we created an installable package to ship to our end users. In the next section, we will learn to deploy on the Android platform.

Deploying on Android

In addition to desktop platforms such as Windows, Linux, and macOS, mobile platforms are equally important due to the massive number of users. Many developers want to make their applications available on mobile platforms. Let's have a look at how that's done. We will briefly discuss deployment considerations on Android.

In *Chapter 5*, *Cross-Platform Development*, you have learned how to create an `.apk` file, which is the deployment package for the Android platform. So, we won't be discussing the steps again. In this section, we will discuss a few necessary changes before uploading to the play store:

1. Create a simple *HelloWorld* application using the Android Kit from the kit selection screen.

2. Change the build mode to **Release** mode.

3. Open the build settings of your project. You will find several options on the screen:

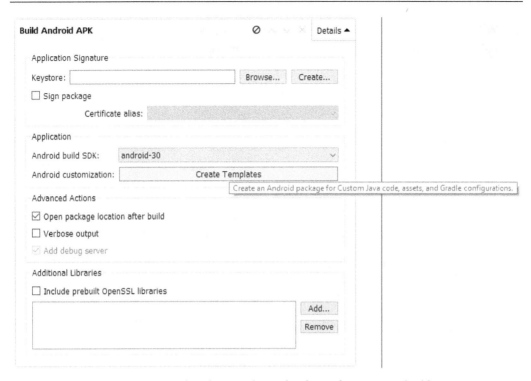

Figure 10.20 – A screenshot showing the Android manifest option in build settings

4. You can see the **Keystore** field under the **Application Signature** section. Click on
    the **Browse…** button to select your existing keystore file or use the **Create…** button
    to create a new keystore file. It can protect key material from unauthorized use. This
    is an optional step and only required for signing your deployment binary.

5. When you click the **Create…** button, then you will see a dialog with several fields. Fill in the related fields and click on the **Save** button. *Figure 10.21* shows the keystore creation dialog:

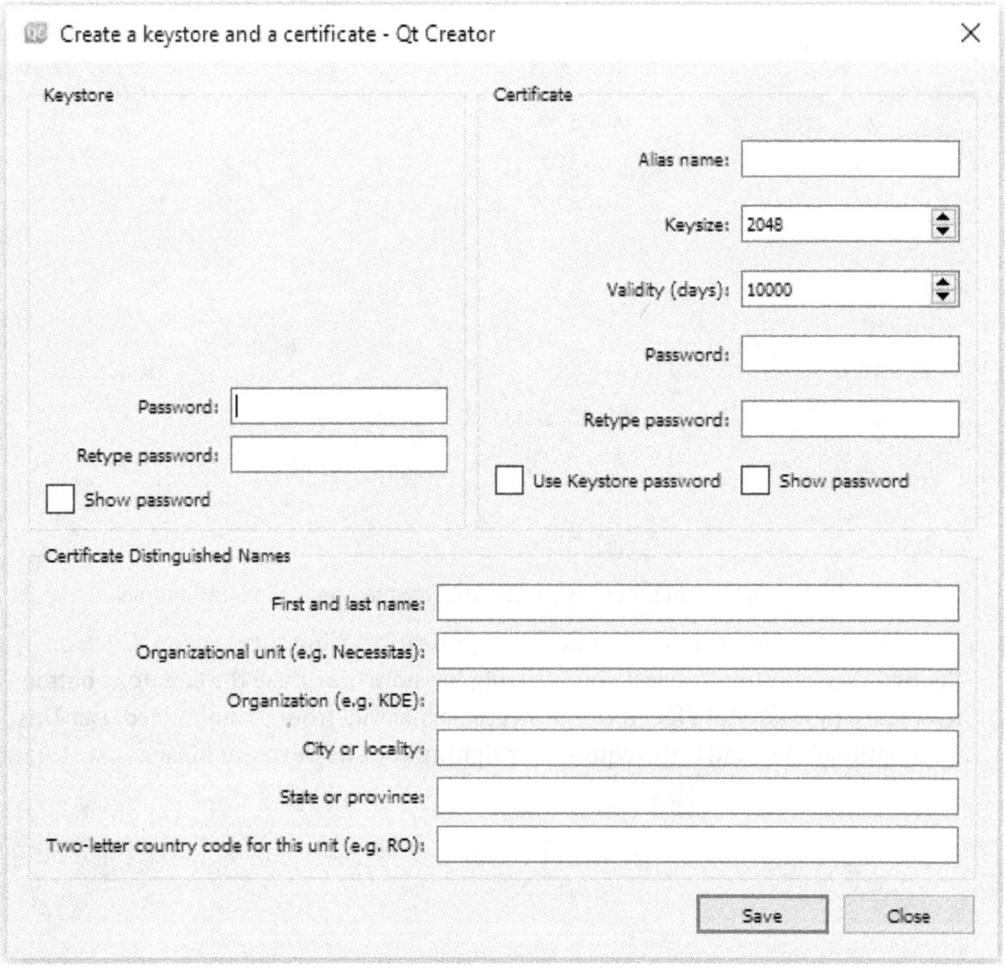

Figure 10.21 – A screenshot showing the keystore creation screen

6. Save the keystore file anywhere, making sure to end the filename with `.keystore`.

The next step is to sign the application package. This is also an optional step and is only required for play store publication. You can learn more about application signing in the official documentation available at `https://developer.android.com/studio/publish/app-signing`.

7. You can select the target Android version and configure your Android app by creating the `AndroidManifest.xml` file with Qt Creator. To do that, click on the **Create Templates** button on the **Build Android APK** screen. You will see a dialog appear, as shown in the following screenshot:

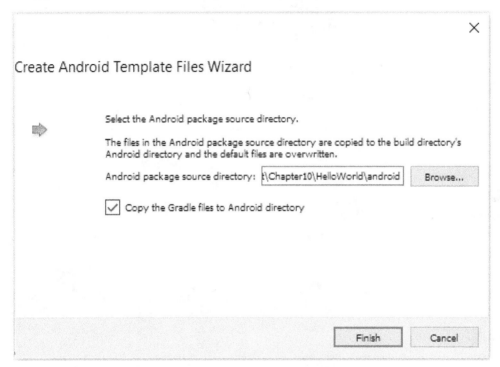

Figure 10.22 – A screenshot showing the manifest file creation wizard

8. Open the manifest file. You will see several options for your Android application.

9. You can set the package name, version code, SDK version, application icon, permissions, and so on. If you add a unique icon, then the default Android icon for your app won't appear on the device. It will make your application unique and easily discoverable on screen.

10. Let's add *HelloWorld* as the application name and add the Qt icon as our application icon, as shown in the following screenshot:

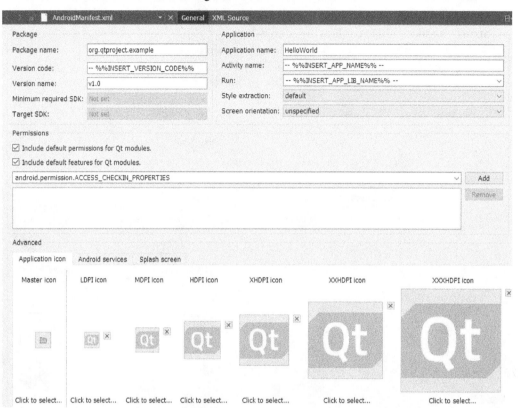

Figure 10.23 – Android manifest file showing different available options

11. Add additional libraries if you are using any third-party libraries such as OpenSSL.

12. Click on the **Run** button in the bottom-left corner of Qt Creator to build and run the application on an Android device. You can also hit the **Deployment** button below the **Run** button to create the deployment binaries.

13. You will see a new dialog appear on the screen. This dialog allows you to choose between your physical Android hardware or the software-emulated virtual device.

14. Connect your Android device and click on the **Refresh Device List** button. Don't forget to enable **Developer options** from your Android device settings. Allow **USB Debugging** when your Android device prompts you:

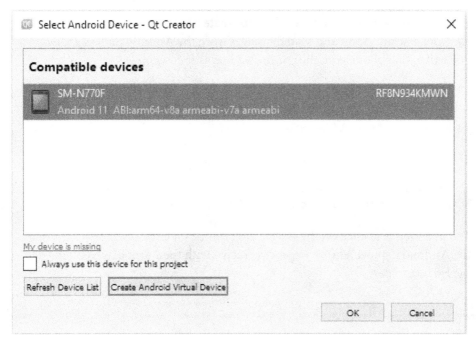

Figure 10.24 – Android device selection dialog

15. If you want to use a virtual device, click on the **Create Android Virtual Device** button. You will see the following screen appear:

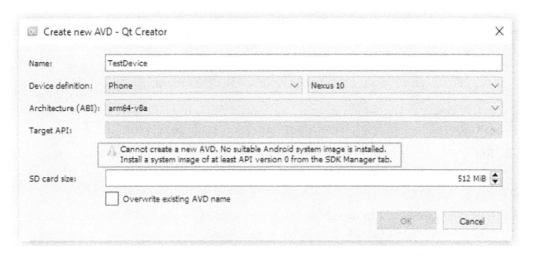

Figure 10.25 – Android virtual device creation screen

16. If the screen warns you about a failure to create a new AVD, then update the Android platform tools and system images from the Android SDK manager. You can update these from the command line as follows:

```
>sdkmanager "platform-tools" "platforms;android-30"
>sdkmanager "system-images;android-30;google_apis;x86"
>sdkmanager --licenses
```

17. Then, run the following command to run avdmanager:

```
>avdmanager create avd -n Android30 -k "system-
images;android-30;google_apis;x86"
```

18. The final step is to click on the **Run** button on the Qt Creator. You will see that the Android deployment package was created with the .apk extension inside the build folder:

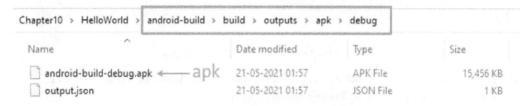

Figure 10.26 – Android installer file generated inside the build directory

19. Internally, Qt runs the androiddeployqt utility. Sometimes, the tool may fail to create the package with the following error:

```
error: aidl.exe ...Failed to GetFullPathName
```

In that case, put your application with a shorter file path and ensure that no directory in your file path has whitespaces. Then, build the application.

20. You can distribute the .apk files to your friends or users. Users have to accept an option saying **Install from Unknown Sources** in their Android mobile or tablets. To avoid this, you should publish your application on the play store.

21. However, if you want to distribute your apps on the Google Play Store, then you have to register as a Google Play developer and sign the package. Google charges a small one-off amount to allow developers to publish their applications.

22. Please note that Qt considers Android apps as a closed source. So, you will require a commercial Qt license if you want to keep your Android app code private.

Congratulations! You have successfully generated a deployable Android application. Unlike iOS, Android is an open system. You can copy or distribute the `.apk` file into other Android devices running on the same Android version and install it.

In this section, we created an installable package for our Android device. In the next section, we will learn about a few more installation tools.

Other installation tools

In this section, we will discuss some additional tools you can use to create an installer. Note that we won't be discussing these tools in detail. I have not verified these installation frameworks with Qt 6. You may visit the respective tool website and learn from their documentation. In addition to the installation framework and tools provided by Qt, you can use the following tools on your Windows machine:

- **CQtDeployer** is an application to extract all the dependent libraries of executables and create a launch script for your application. The tool claims to deploy applications faster and provides flexible infrastructure. It supports both Windows and Linux platforms. You can learn more about this tool at the following link: `https://github.com/QuasarApp/CQtDeployer`.

- **Nullsoft Scriptable Install System** (**NSIS**) is a script-driven installation authoring tool from Nullsoft, the same company that created Winamp. It has become a popular alternative to proprietary commercial tools such as InstallShield. The current version of NSIS has a modern graphical user interface, LZMA compression, multilingual support, and a simple plugin system. You can explore more about the tool at `https://nsis.sourceforge.io/Main_Page`.

- **InstallShield** is a proprietary software application that allows you to create installers and software bundles. InstallShield is generally used to install the software on Windows Platform desktop and server systems, but it may also be used to manage software applications and packages on a wide range of portable and mobile devices. Have a look at its features and play with the trial version. You can download the trial and read more about it at `https://www.revenera.com/install/products/installshield.html`.

- **Inno Setup** is a free software script-driven installation system created in Delphi. It was first released in 1997, yet still competes with many commercial installers thanks to its excellent feature set and stability. Learn more about this installer at the following link: `https://jrsoftware.org/isinfo.php`.

You can select any of the installation frameworks and deploy your application. In the end, it should meet your installation goal.

In this section, we discussed a few more installation tools that may be beneficial for your needs. Let's now summarize our takeaways from this chapter.

Summary

We began the chapter by discussing the application deployment problem and learning the difference between static and dynamic libraries. Then we discussed the different deployment tools in Qt and the specific case of Windows deployment and installation. Armed with this knowledge, we deployed a sample app on Windows and created an installer using the Qt Installer Framework. In addition, we discovered deploying applications on Linux and macOS and honed our skills for deploying applications on various platforms. Later, we explained some of the important points to be considered before publishing a Qt-based Android application to the play store.

Finally, we looked at some third-party installer utilities. To summarize, you have learned to develop, test, and deploy a Qt application on various platforms. With this knowledge, you should be able to create your installation packages and share them with the world.

In *Chapter 11*, *Internationalization*, we will learn about developing a translation-aware Qt application.

11
Internationalization

In earlier chapters, we learned how to create GUI applications with Qt Widgets or Qt Quick. To make our applications usable across the world, we need to add translations to the application.

The process of making your application translation-aware is known as **internationalization**. It makes it easy to localize content for viewers from different cultures, regions, or languages. Translating Qt Widgets and Qt Quick apps into local languages is very easy with Qt. These processes of adapting an application to different languages with the geographical and technical standards of a target market are known as **internationalization**.

You will learn how to make an application with multilingual support. Throughout the chapter, we will explore different tools and processes to make a translation-aware application. In this chapter, we will discuss the following:

- Basics of internationalization
- Writing source code for translation
- Loading translation files
- Internationalization with Qt Widgets
- Internationalization with Qt Quick
- Deploying translations

By the end of this chapter, you will be able to create a translation-aware application using Qt Widgets and Qt Quick.

Technical requirements

The technical requirements for this chapter include minimum versions of Qt 6.0.0 and Qt Creator 4.14.0 installed on the latest desktop platform such as Windows 10, Ubuntu 20.04, or macOS 10.14.

All the code used in this chapter can be downloaded from the following GitHub link: `https://github.com/PacktPublishing/Cross-Platform-Development-with-Qt-6-and-Modern-Cpp/tree/master/Chapter11`.

> **Important note**
> The screenshots used in this chapter are taken on the Windows platform. You will see similar screens based on the underlying platform on your machine.

Understanding internationalization and Qt Linguist

The processes of adjusting an application to different languages, geographical variations, and technological specifications of a target market are known as **internationalization** and **localization**. Internationalization refers to the process of creating a software application that can be translated into a variety of languages and for different regions without requiring significant technical changes. Internationalization is often abbreviated to **i18n**, with 18 being the number of letters between the letters *i* and *n* in the English word. The ease with which a product can be localized is greatly influenced by its internationalization. Creating a linguistically and culturally focused application for a global market is a much more complex and time-consuming process. Hence, companies focus on creating i18n-aware applications for global markets from the beginning of product development.

For internationalization, you should design your application in such a manner that it avoids roadblocks for localization or global deployment later. This covers aspects such as allowing Unicode or maintaining careful handling of legacy character encodings where appropriate, taking caution of string concatenation, preventing code dependencies on user interface string values, and so on. You should provide support for features such as identifying translatable strings and system language that may be required for internationalization later.

Your application should be aware of local languages, date and time formats, numeral systems, or cultural preferences. The modification of a product, application, or document's content to fulfill the language, cultural, and other preferences of a particular target market is known as **localization**. Localization is often written in English as **l10n**, where 10 is the number of letters between *l* and *n*. Localization entails incorporating region-specific requirements and translating applications for a specific region or language. Localizable features should be separated from the source code, allowing adaptation as per the user's cultural preferences.

Qt Linguist is a tool that enables users to create translations of your Qt applications. Qt Linguist can be launched from the installation directory or the IDE. The tool comes with two integrated programs, known as `lupdate` and `lrelease`. These programs can be used with a qmake project or directly with the filesystem.

The `lupdate` tool locates translatable strings in the project's source, header, and `.ui` or `.qml` files. Then it creates or updates the translation files (`.ts` files). You can specify the files to be processed on the command line or in a `.pro` file as arguments. `.ts` files use **Document Type Definition (DTD)** format, described at the following link:

`https://doc.qt.io/qt-6/linguist-ts-file-format.html`

Qt provides excellent support for internationalization. Qt has built-in support for many languages in all user interface elements. However, when writing source code for your application, you have to follow certain practices. This includes marking translatable strings, avoiding ambiguous strings, using numbered arguments (`%n`) as placeholders, and loading the right translation file. You can use both C++ and user interface files, and you can also have translatable strings in both sources. The tool locates and adds the strings from all sources into a single `.ts` file with corresponding contexts.

The translation files with `.ts` extension are used during application development. These files can be compiled into a compact binary format. The compiled translation files are encoded in the QM format and have the `.qm` file extension. While running an application, the Qt runtime makes use of `.qm` files instead of `.ts` files. You can convert `.ts` to `.qm` files using the `lrelease` tool. A `.qm` file is a lightweight binary file. It allows lightning-fast translation lookups. You can specify `.ts` files on the command line or in a `.pro` project file to be processed by `lrelease`. This tool is used every time an application is released, from the test version to the final production version. If the `.qm` files aren't available, then the application will still work fine and use the original texts from the source files.

For the selection of languages, Qt Linguist and `lrelease` use certain internal rules. You can find details about these rules at the following link:

`https://doc.qt.io/qt-6/i18n-plural-rules.html`

Let's have a look at the Qt Linguist user interface. You can launch Qt Linguist from the Qt installation directory by double-clicking the **Linguist** executable or selecting it from the command prompt. You will see the following user interface appear on your screen:

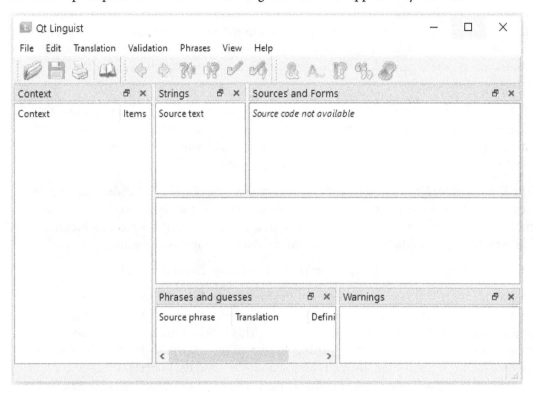

Figure 11.1 – Qt Linguist user interface

In the preceding figure, you can see multiple sections and there are a few disabled buttons in the toolbar. You can open a `.ts` file from the **File** menu. We will discuss these sections while discussing an example in a later section of this chapter.

You can learn more about Qt Linguist and the GUI interface at the following web page:

`https://doc.qt.io/qt-6/linguist-translators.html`

In this section, you got familiar with terms related to internationalization and tools provided by the Qt framework. With a good understanding of the basics, we are ready to write a translation-aware application in the next section.

Writing source code for translation

In this section, we will discuss how to mark strings as translatable strings and how to use the tools provided by Qt. Wherever your application uses a quoted string that is visible to the user, make sure the `QCoreApplication::translate()` method processes it. To do this, simply use the `tr()` method to mark the strings as translatable that are meant for display purposes. This feature is used to show which text strings are translatable inside your C++ source files.

For example, if you want to use a `QLabel` to show text on a user interface, then embed the text inside the `tr()` method as follows:

```
QLabel *label = new QLabel(tr("Welcome"));
```

The class name is the translation context for the `QObject` and its derived classes. To override the context, `QObject`-derived classes must use the `Q_OBJECT` macro in their class definition. This macro sets the context for the derived classes.

Qt provides several convenience macros and methods for internationalization. A few of the most common macros used for translation are as follows:

- `tr()` returns a translated string if translation is available in a C++ source file.
- `qsTr()` returns a translated string if translation is available in a QML file.
- `qtTrId()` finds and returns a translated string identified by an ID in a C++ file.
- `qsTrId()` finds and returns a translated string identified by an ID in a QML file.
- `QT_TR_NOOP()` tells `lupdate` to collect the string in the current context for translating later.
- `QT_TRID_NOOP()` marks an ID for dynamic translation.
- `QCoreApplication::translate()` provides the translation by querying the installed translation files.
- `qsTranslate()` provides a translated version for a given context in a QML file.
- `QQmlEngine::retranslate()` updates all binding expressions with strings marked for translation.

Translatable strings in C++ files are marked using `tr()`, and in QML files `qsTr()` is used. We will discuss these macros and methods throughout this chapter.

All the translatable strings are fetched by the `lupdate` tool and updated in a **Translation Source (TS)**. A TS file is an XML file. Usually, TS files follow the following naming convention:

```
ApplicationName>_<LanguageCode>_<CountryCode>.ts
```

In this convention, `LanguageCode` is an ISO 639 language code in lowercase and `CountryCode` is an ISO 3166 two-letter country code in uppercase. You can create translations for the same language but targeting a different country by using the specific country code. You can create a default translation file with a language code and country code while creating a Qt application through Qt Creator's new project wizard.

Once you create the `.ts` files, you can run `lupdate` to update the `.ts` files with all the user-visible strings. You can run `lupdate` from the command line as well as from Qt Creator and the Visual Studio add-in. Let's use Qt's command prompt to run the following command for the `HelloWorld` application:

```
>lupdate HelloWorld.pro
```

`lupdate` fetches the translatable strings from different source files such as `.cpp`, `.h`, `.qml`, and `.ui`. For `lupdate` to work effectively, you should specify the translation files in the application's `.pro` file under the `TRANSLATIONS` variable. Look at the following `.pro` file section where we have added six translation source files:

```
TRANSLATIONS = \
        HelloWorld_de_DE.ts \
        HelloWorld_fi_FI \
        HelloWorld_es_ES.ts \
        HelloWorld_zh_CN.ts \
        HelloWorld_zh_TW.ts \
        HelloWorld_ru_RU.ts
```

You can also add wildcard-based translation file selections with `*.ts`.

To translate a Qt Quick application, use the `qsTr()` method to mark the strings inside the `.qml` files. You can create a translation file for a single QML file as follows:

```
>lupdate main.qml -ts HelloWorld_de_DE.ts
```

You can create multiple translation files for different languages and put them inside a `.qrc` file:

```
RESOURCES += translations.qrc
```

You can process all QML files in a `.qrc` file using `lupdate` as follows:

```
>lupdate qml.qrc -ts HelloWorld_de_DE.ts
```

To process all QML files without using a `.qrc` file, type the following into Qt's command prompt:

```
>lupdate -extensions qml -ts HelloWorld_de_DE.ts
```

You can also pass a C++ source file as an argument along with the resource file. It is optional to mention translation files in the `.pro` file. You can do it by specifying the translation file on the command line as follows:

```
>lupdate qml.qrc messages.cpp -ts HelloWorld_de_DE.ts
 HelloWorld _es_ES.ts
```

`lrelease` integrates translations that are marked as `finished`. If a string is missing translations and is marked as `unfinished`, then the original text is used. Translators or developers can modify the TS file contents and mark it as `finished` with the following steps:

1. Launch Qt Linguist and open the `.ts` file from the **File** menu. Alternatively, right-click on the `.ts` file in the project structure and open with Qt Linguist, as shown here:

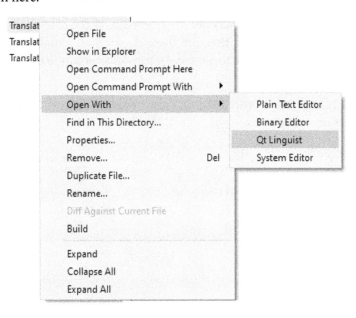

Figure 11.2 – The Open With Qt Linguist option in Qt Creator

2. Then click on any of the contexts in the **Context** view to see the translatable strings for that context in the **Strings** view.

3. In the **Source text** view, enter the translation of the current string. You can find existing translations and similar phrases in the **Phrases and Guesses** view.

4. The translator may enter a comment in the **Translator comments** field.

5. To finish the translation, press *Ctrl + Enter* and select the tick icon from the toolbar. You will see a green tick mark for translated strings.

6. Finally, save the file and exit the Qt Linguist tool.

You can run `lrelease` without specifying a `.pro` file. When you run `lrelease` to read the `.ts` files, then it generates `.qm` files that are used by the application at runtime:

```
>lrelease *.ts
```

Once the `.qm` files are generated, add them to a `.qrc` file. Your application is now ready for translation.

You can also use a text ID-based translation mechanism. In this approach, every translatable string in the application is assigned a unique identifier. These unique text identifiers are directly used in the source code as a replacement for actual strings. The user interface developers would need to put more effort into this but it is much easier to maintain if your application contains huge numbers of translated strings.

In some applications, certain classes may not use `QObject` as the base class or use the `Q_OBJECT` macro in their class definition. But these classes may contain some strings that may require translation. To solve this issue, Qt provides certain macros to add translation support.

You can use `Q_DECLARE_TR_FUNCTIONS(ClassName)` as follows to enable translation for your non-Qt classes:

```
class CustomClass
{
    Q_DECLARE_TR_FUNCTIONS(CustomClass)
public:
    CustomClass();
    ...
};
```

This macro is defined inside `qcoreapplication.h`. When you add this macro, Qt adds the following function to your class to enable translation:

```
static inline QString tr(const char *sourceString, const char
*disambiguation = nullptr, int n = -1)
{
    return QCoreApplication::translate(#className,
sourceString,
disambiguation, n);
}
```

From the preceding code, you can notice that Qt calls `QCoreApplication::translate()` with the class name as the context.

You can also have translatable strings outside a class or method; `QT_TR_NOOP()` and `QT_TRANSLATE_NOOP()` are used to mark these strings as translatable. There are different macros and functions available for text ID-based translation. You can use `qsTrId()` instead of `qsTr()`, and `QT_TRID_NOOP()` instead of `QT_TR_NOOP()`. You can use the same text IDs as user interface strings instead of plain strings in the user interface.

In Qt Linguist, multiple translation files can be loaded and edited simultaneously. You can also use **phrase books** to reuse existing translations. Phrase books are standard XML files that contain typical phrases and their translations. These files are created and updated by Qt Linguist and can be used by any number of projects and applications. If you would like to translate source strings that are available in a phrase book, Qt Linguist's Batch Translation function can be used. Select **Batch Translation** to specify which phrase books to use and in what order during the batch translation process. Only entries with no current translation should be considered, and batch-translated entries should be marked as **Accepted**. You can also create a new phrase book from the **New Phrase Book** option.

> **Important note**
>
> `lupdate` requires all source code to be encoded in UTF-8 by default. Files that feature a **Byte Order Mark (BOM)** can also be encoded in UTF-16 or UTF-32. You have to set the CODECFORSRC qmake variable to UTF-16 to parse files without a BOM as UTF-16. By default, certain editors such as Visual Studio use a separate encoding. You can avoid encoding problems by limiting source code to ASCII and using escape sequences for translatable strings.

In this section, we discussed how to use `lupdate` and `lrelease` to create and update translation files. Next, we will learn how to install a translator and load a translation file in your Qt application.

Loading translations in a Qt application

In the previous section, we created translation files and understood the uses of the tools. To look up translations in a TS file, QTranslator functions are used. The translator must be instantiated before the application's GUI objects.

Let's have a look at how to load these translation files using QTranslator in the following code snippet:

```
QTranslator translator;
if(translator.load(QLocale(),QLatin1String("MyApplication")
            , QLatin1String("_"), QLatin1String(":/i18n")))
    {
        application.installTranslator(&translator);
    }
    else
    {
        qDebug() << "Failed to load. "
                << QLocale::system().name();
    }
```

In the preceding code, you can see that we have created a `translator` object and loaded the corresponding translation file. QLocale is used to fetch the underlying system language. You can also use QLocale for localizing numbers, the date, the time, and currency strings.

Alternatively, you can load a translation file as follows:

```
QString fileName = ":/i18n/MyApplication_"+QLocale::
                system().name()
+".qm";
translator.load(fileName);
```

Here, we are looking into the system language and loading the corresponding translation files. The preceding approach works well when you want to use the system language as your application language. However, some users may like to use a regional language that is different from the system language. In that case, we can change the language as per user choice. We will learn how to do that in the next section.

Switching languages dynamically

So far, you have learned how to use the system language or a default language for your Qt application. In most applications, you can just detect the language in `main()` and load an appropriate `.qm` file. Sometimes, your application must be able to support changes to the user's language settings while still running. An application that is used by multiple people in shifts may need to switch languages without requiring a restart.

To achieve this in a Qt Widgets-based application, you can override `QWidget::changeEvent()`. Then, you have to check whether the event is of the `QEvent::LanguageChange` type. You can retranslate the user interface accordingly.

The following code snippet explains how to achieve dynamic translation in a Qt Widgets-based GUI:

```
void CustomWidget::changeEvent(QEvent *event)
{

    if (QEvent::LanguageChange == event->type())
    {

        ui->retranslateUi(this);

    }
    QWidget::changeEvent(event);

}
```

`QEvent::LocaleChange` can cause the list of installed translators to switch. You can create an application with a user interface that provides users with the option to change the current application language. When the `QEvent::LanguageChange` event occurs, the default event handler for `QWidget` subclasses will call this method. If you are using the `QCoreApplication::installTranslator()` function to install a new translation, you will get a `LanguageChange` event. In addition, by sending `LanguageChange` events to other widgets, the GUI will force them to update. Any other events can be passed to the base class for further processing.

To enable dynamic translation, you can provide an option in the command line or over a GUI. By default, Qt puts all of the translatable strings in the `.ui` file inside `retranslateUi()`. You have to call this function whenever the language is changed. You can also create and call your custom method to retranslate the strings created through C++ code based on the `QEvent::LanguageChange` event.

In this section, we discussed how to achieve dynamic translation in the application runtime. In the next section, we will create a translation-aware application using Qt Widgets.

Internationalization with Qt Widgets

In the previous sections, we discussed how to create translation files and how to use QTranslator to load a translation file. Let's create a simple example using Qt Widgets and implement our learning.

Follow the subsequent steps to create the sample application:

1. Create a Qt Widgets-based application using Qt Creator's new project creation wizard and follow through the screens as discussed in earlier chapters.

2. On the **Translation File** screen, choose **German (Germany)** as the language option, or any preferred language.

3. Finish the project creation. You will see that Simplei18nDemo_de_DE.ts is created in your project structure.

4. Next, you add a QLabel to the .ui file and add Welcome text.

5. Next, run lupdate. You can run lupdate from the command line as well as from the Qt Creator interface, as shown in *Figure 11.3*:

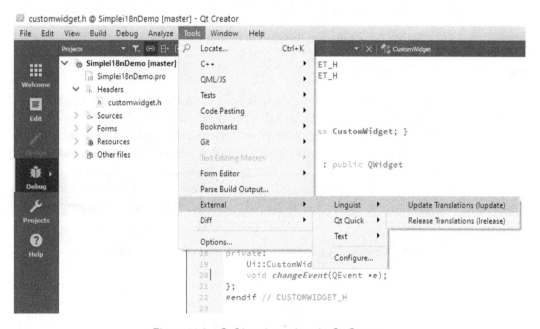

Figure 11.3 – Qt Linguist options in Qt Creator

6. When you run `lupdate`, you will see the following output in the console window:

```
C:\Qt6Book\Chapter11\Simplei18nDemo>lupdate
Simplei18nDemo.pro
Info: creating stash file C:\Qt6Book\Chapter11\
Simplei18nDemo\.qmake.stash
Updating 'Simplei18nDemo_de_DE.ts'...
    Found 2 source text(s) (2 new and 0 already existing)
```

7. Now, the `.ts` file is updated with strings. Open `Simplei18nDemo_de_DE.ts` with a plain text editor. You should see the following content:

```xml
<?xml version="1.0" encoding="utf-8"?>
<!DOCTYPE TS>
<TS version="2.1" language="de_DE">
<context>
    <name>CustomWidget</name>
    <message>
        <location filename="customwidget.ui"
            line="14"/>
        <source>Simple i18n Demo</source>
        <translation type="unfinished"></translation>
    </message>
    <message>
        <location filename="customwidget.ui"
            line="25"/>
        <source>Welcome</source>
        <translation type="unfinished"></translation>
    </message>
</context>
</TS>
```

You can see that the user interface strings are updated inside the .ts file and the language is defined for the translation at the top of the file. You can create respective translation files by modifying this field in the code:

```
<TS version="2.1" language="de_DE">
```

You will also see that the translation status is `unfinished`.

8.   So, let's open the file with Qt Linguist and complete the translation:

Figure 11.4 – Example showing different sections of the Qt Linguist interface

9.   You will see six different sections in the user interface. Select a context in the **Context** view to load the corresponding strings.

10.   Add a translation in the **Source text** view. You can use Google Translate to translate the string into the desired language. Here we have translated the strings to the German language using Google Translate.

> **Note**
>
> There are multiple translations used. Please ignore if the strings don't have the exact meaning. I am not familiar with the German language. I have used this for demonstration purposes. Hence, I have added a translator's comments.

11. To complete the translation, press *Ctrl + Enter* or click on the green tick icon on the toolbar.

12. The next step is to save the translation. Repeat this for all translatable strings listed in the context.

13. Run `lrelease` from Qt's command prompt or the IDE's option. You will see that the `.qm` file is generated:

```
C:\Qt6Book\Chapter11\Simplei18nDemo>lrelease *.ts
Updating 'Simplei18nDemo_de_DE.qm'...
    Generated 2 translation(s) (2 finished and 0
unfinished)
```

14. Let's add the translator to `main.cpp` and load the translation file:

```cpp
#include "customwidget.h"
#include <QApplication>
#include <QTranslator>
#include <QDebug>
int main(int argc, char *argv[])
{
    QApplication app(argc, argv);
    QTranslator translator;
    if(translator.load(":/translations
                        /Simplei18nDemo_de_DE.qm"))
    {
        app.installTranslator(&translator);
        qDebug()<<"Loaded successfully!";
    }
    else
    {
        qWarning()<<"Loading failed.";
    }
    CustomWidget customUI;
    customUI.show();
    return app.exec();
}
```

15. The final step is to run qmake and build the application. Then, hit the **Run** button in the bottom-left corner.

16. We have successfully translated our GUI into German. You will see the following output:

Figure 11.5 – Output of the translation example using Qt Widgets

Congratulations! You learned how to translate your application into a different language. You can now translate your Qt application to a local language and share it with your friends and colleagues.

In this section, we discussed how to create a translation-aware application using Qt Widgets. In the next section, we will add dynamic translation capability to the Qt Widgets application.

Adding dynamic translation to a Qt Widgets application

In the previous section, you learned how to create a Qt Widgets-based application and change the language to a preferred language. However, like most global applications, you may need to provide more translations and allow users to change the language on the fly.

Let's modify the example from the preceding section with some additional implementations:

1. Add a combo box to the `.ui` file and add three languages to it. For explanation purposes, we have used English, German, and Spanish. We have added a message at the center and added a language-switching option in a dropdown:

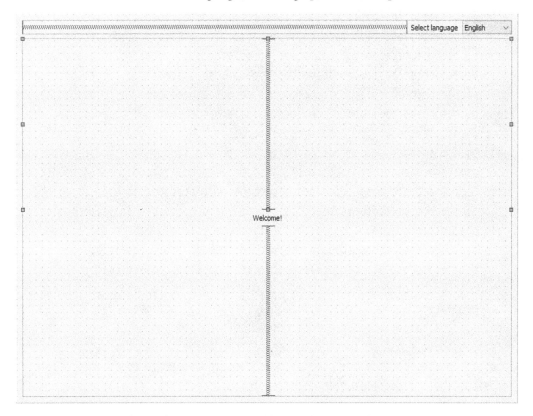

Figure 11.6 – A form in Qt Designer showing layouts used in the example

2. Add the new translation files to the project file as follows:

```
TRANSLATIONS += \
    WidgetTranslationDemo_en_US.ts \
    WidgetTranslationDemo_de_DE.ts \
    WidgetTranslationDemo_es_ES.ts
```

3. Let's modify the `CustomWidget` class and add the following methods for dynamic translation:

```cpp
#ifndef CUSTOMWIDGET_H
#define CUSTOMWIDGET_H
#include <QWidget>
#include <QTranslator>
QT_BEGIN_NAMESPACE
namespace Ui { class CustomWidget; }
QT_END_NAMESPACE
class CustomWidget : public QWidget
{
    Q_OBJECT
public:
    CustomWidget(QWidget *parent = nullptr);
    ~CustomWidget();
  public slots:
    void languageChanged(int index);
    void switchTranslator(const QString& filename);
    void changeEvent(QEvent *event);
private:
    Ui::CustomWidget *ui;
    QTranslator m_translator;
};
#endif // CUSTOMWIDGET_H
```

4. The next step is to connect the signal and slot. We have created the connections in the constructor:

```cpp
CustomWidget::CustomWidget(QWidget *parent)
    : QWidget(parent), ui(new Ui::CustomWidget)
{
    ui->setupUi(this);
    connect(ui->languageSelectorCmbBox,
            SIGNAL(currentIndexChanged(int)),this,
            SLOT(languageChanged(int)));
    qApp->installTranslator(&m_translator);
}
```

5. Let's add the following code to the slot definition:

```
void CustomWidget::languageChanged(int index)
{
    switch(index)
    {
    case 0: //English
        switchTranslator(":/i18n/
            WidgetTranslationDemo_en_US.qm");
        break;
    case 1: //German
        switchTranslator(":/i18n/
            WidgetTranslationDemo_de_DE.qm");
        break;
    case 2: //Spanish
        switchTranslator(":/i18n/
            WidgetTranslationDemo_es_ES.qm");
        break;
    }
}
```

Here, we are receiving the language choice from the user interface through the combo box index change signal.

6. The next step is to install a new translator:

```
void CustomWidget::switchTranslator(const QString&
filename)
{
    qApp->removeTranslator(&m_translator);
    if(m_translator.load(filename))
    {
        qApp->installTranslator(&m_translator);
    }
}
```

7. The last step is to reimplement `changeEvent()`:

```
void CustomWidget::changeEvent(QEvent *event)
{
    if (event->type() == QEvent::LanguageChange)
    {
        ui->retranslateUi(this);
    }
    QWidget::changeEvent(event);
}
```

8. Run qmake and hit the **Run** button on the IDE.

    The following screen will appear:

Figure 11.7 – Example showing the output when the English language is selected

9. Change the language from the language selection dropdown. Let's select **German** as the new language. You will see the entire GUI changed with German strings:

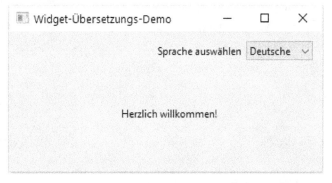

Figure 11.8 – Example showing the output when the German language is selected

10. Again, switch the language to **Spanish**. You will see the GUI text changed to Spanish:

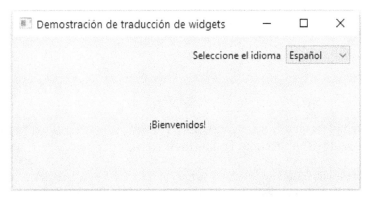

Figure 11.9 – Example showing the output when the Spanish language is selected

Congratulations! You have successfully created a multilingual Qt Widgets application.

In this section, you learned how to translate your Qt Widgets-based GUI at runtime. In the next section, we will create a translation-aware application using Qt Quick.

Internationalization with Qt Quick

In the previous section, we discussed internationalization in Qt Widgets. In this section, we will discuss different aspects of internationalizing your Qt Quick application. The underlying localization scheme in Qt Quick applications is similar to Qt Widgets applications. The same set of tools described in the Qt Linguist Manual are also used in Qt Quick. You can translate an application that uses both C++ and QML.

In a Qt project file, the SOURCES variable is used for C++ source files. If you list QML or JavaScript files under this variable, the compiler will attempt to use the files considering them as C++ files. As a workaround, you can use a lupdate_only {...} conditional declaration to make the QML files visible to the lupdate tool but invisible to the C++ compiler.

Consider the following example. The application's .pro file snippet lists two QML files:

```
lupdate_only {
SOURCES = main.qml \
          HomeScreen.qml
}
```

You may also use a wildcard match to specify the QML source files. Since the search is not recursive, you must list each directory in which user interface strings can be found in the source code:

```
lupdate_only{
SOURCES = *.qml \
          *.js
}
```

Let's create an example with a simple translation. We will create a similar screen as we created in the Qt Widgets application. Follow these steps:

1. Create a Qt Quick-based application using Qt Creator's new project creation wizard and follow through the screens as discussed in earlier chapters.

2. On the **Translation File** screen, choose **German (Germany)** as the language option or any preferred language.

3. Finish the project creation. You will see that QMLTranslationDemo_de_DE.ts is created in your project structure.

4. Next, you add a Text to the .qml file and add Welcome text:

```
import QtQuick
import QtQuick.Window
Window {
    width: 512
    height: 512
    visible: true
    title: qsTr("QML Translation Demo")
    Text {
        id: textElement
        anchors.centerIn: parent
        text: qsTr("Welcome")
    }
}
```

5. Add the following lines of code to main.cpp:

```
#include <QGuiApplication>
#include <QQmlApplicationEngine>
#include <QTranslator>
```

```cpp
#include <QDebug>
int main(int argc, char *argv[])
{
    QGuiApplication app(argc, argv);
    QTranslator translator;
    if(translator.load(":/translations/
        QMLTranslationDemo_de_DE.qm"))
    {
        app.installTranslator(&translator);
        qDebug()<<"Loaded successfully!";
    }
    else
    {
        qWarning()<<"Loading failed.";
    }
    QQmlApplicationEngine engine;
    const QUrl url(QStringLiteral("qrc:/main.qml"));
    QObject::connect(&engine,
        &QQmlApplicationEngine::objectCreated,
            &app, [url](QObject *obj, const QUrl
                &objUrl)
            {
                if (!obj && url == objUrl)
                    QCoreApplication::exit(-1);
            }, Qt::QueuedConnection);
    engine.load(url);
    return app.exec();
}
```

6. The steps are similar to the Qt Widgets example. Next, run `lupdate`.

7. Follow the same steps to update the translation in the `.ts` files using Qt Linguist.

8. Run `lrelease` from Qt's command prompt or from the IDE's option. You will see that the `.qm` file is generated.

9. Add the `.qm` files to the resources (`.qrc`) file and run qmake.

10. The last step is to build and run the application. Hit the **Run** button in Qt Creator.

11. You will see the identical output as we have seen in the Qt Widgets example:

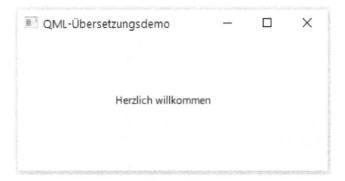

Figure 11.10 – Output of the translation example using Qt Quick

In the preceding example, we translated our Qt Quick application to German.

In this section, we discussed how to create a translation-aware application using Qt Quick. In the next section, we will add dynamic translation capability to the Qt Quick application.

Translating dynamically in a Qt Quick application

In the previous section, you learned how to create a Qt Quick-based application and how to change the language to a preferred language. Just like the Qt Widgets example, you can also add dynamic translations to your Qt Quick application.

Let's modify the previous example with some additional implementations:

1. Create an i18n support class named `TranslationSupport` and add the following lines:

```
#ifndef TRANSLATIONSUPPORT_H
#define TRANSLATIONSUPPORT_H
#include <QObject>
#include <QTranslator>
class TranslationSupport : public QObject
{
    Q_OBJECT
public:
```

```
        explicit TranslationSupport(QObject *parent =
                                    nullptr);
    public slots:
        void languageChanged(int index);
        void switchTranslator(const QString& filename);
    signals:
        void updateGUI();
    private:
        QTranslator m_translator;
    };
    #endif // TRANSLATIONSUPPORT_H
```

The preceding code is a helper class that supports the translation feature in QML. It is used to update the translation files in the translator.

2. In the next step, add the following code to switch the translator:

```
    void TranslationSupport::switchTranslator(const QString&
    filename)
    {
        qApp->removeTranslator(&m_translator);
        if(m_translator.load(filename))
        {
            qApp->installTranslator(&m_translator);
            emit updateGUI();
        }
    }
```

3. Then, add the following code to the QML INVOKABLE method definition:

```
    void TranslationSupport::languageChanged(int index)
    {
        switch(index)
        {
        case 0: //English
            switchTranslator(":/i18n/
                QMLDynamicTranslation_en_US.qm");
            break;
        case 1: //German
```

```
        switchTranslator(":/i18n/
            QMLDynamicTranslation_de_DE.qm");
        break;
    case 2: //Spanish
        switchTranslator(":/i18n/
            QMLDynamicTranslation_es_ES.qm");
        break;
    }
}
```

4. In the `main.cpp` file, add the following code. Please note that we have exposed the `TranslationSupport` instance to the QML engine:

```
#include <QGuiApplication>
#include <QQmlApplicationEngine>
#include <QQmlContext>
#include "translationsupport.h"
int main(int argc, char *argv[])
{
    QGuiApplication app(argc, argv);
    TranslationSupport i18nSupport;
    QQmlApplicationEngine engine;
    engine.rootContext()->setContextProperty(
        "i18nSupport", &i18nSupport);
    const QUrl url(QStringLiteral("qrc:/main.qml"));
    QObject::connect(&i18nSupport,
        &TranslationSupport::updateGUI, &engine,
        &QQmlApplicationEngine::retranslate);
    engine.load(url);
    return app.exec();
}
```

5. Then add the `updateGUI()` signal with the `QQmlApplicationEngine::retranslate()` method.

6. Let's have a look at the `main.qml` file. We have added a combo box to the `.qml` file and added three languages to it. For explanation purposes, we have used English, German, and Spanish:

```qml
Text {
    id: textElement
    anchors.centerIn: parent
    text: qsTr("Welcome!")
}
Row {
    anchors {
        top: parent.top;      topMargin: 10 ;
        right: parent.right;  rightMargin: 10;
    }
    spacing: 10
    Text{
        text: qsTr("Select language")
        verticalAlignment: Text.AlignVCenter
        height: 20
    }
    ComboBox {
        height: 20
        model: ListModel {
            id: model
            ListElement { text: qsTr("English") }
            ListElement { text: qsTr("German") }
            ListElement { text: qsTr("Spanish") }
        }
        onCurrentIndexChanged: {
            i18nSupport.languageChanged(currentIndex)
        }
    }
}
```

7. Run `lupdate` and proceed with the translation process.

8. Follow the same steps to update the translation in `.ts` files using Qt Linguist.

9. Run `lrelease` from Qt's command prompt or from the IDE's option. You will see that the .qm file is generated.

10. Add the .qm files to the resources (.qrc) file and run qmake.

11. The last step is to build and run the application. Hit the **Run** button in Qt Creator.

    The following screen will appear:

Figure 11.11 – Qt Quick example showing the output when the English language is selected

12. Change the language from the language selection dropdown. Let's select **German** as the new language. You will see the entire GUI changed with German strings:

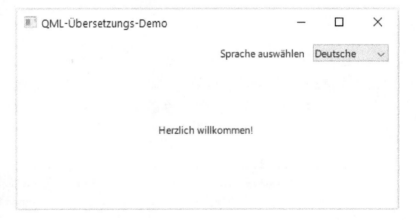

Figure 11.12 – Qt Quick example showing the output when the German language is selected

13. Again, switch the language to **Spanish**. You will see the GUI text changed to Spanish:

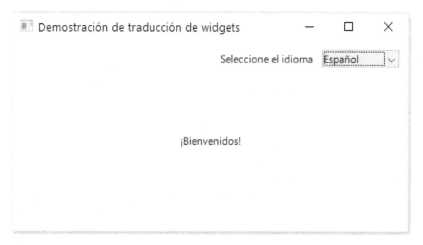

Figure 11.13 – Qt Quick example showing the output when the Spanish language is selected

Congratulations! You have successfully created a multilingual Qt Quick application.

In this section, you learned how to translate your Qt Quick-based GUI at runtime. In the next section, we will discuss how to deploy translation files.

Deploying translations

In previous sections, we learned how to create translation-aware applications using both Qt Widgets and QML. You don't have to ship the .ts files with your application. To deploy translations, your release team must use the updated .qm files and ship them with the application package. The .qm files required for the application should be placed in a location where QTranslator can locate them. Typically, this is done by embedding qm files in a resource (.qrc) file or specifying a path that contain the .qm files relative to QCoreApplication::applicationDirPath(). The rcc tool is used to embed the translation files into a Qt application during the build process. It works by producing a corresponding C++ file containing specified data.

You can automate the generation of .qm files by adding a script to your .pro file. You do it by following these steps:

1. To begin, use the language codes to declare the languages under the LANGUAGES variable in your Qt project (.pro) file.

2. Add lrelease and embed_translations to the CONFIG variable.

3. Then add a function to generate the .ts files for the intended languages.

4. Finally, define the TRANSLATIONS_FILES variable, use lrelease to create the .qm files, and embed them in the application resources.

The preceding steps will run the lrelease automatically and generate the .qm files. The lrelease tool processes the translation files listed under the TRANSLATIONS and EXTRA_TRANSLATIONS. Unlike the TRANSLATIONS variable, files listed under EXTRA_TRANSLATIONS are only processed by lrelease tool, not by the lupdate. You need to embed the .qm files into your resource or ship the .qm files with your deployment package.

You can learn more about automating generation of QM files here: https://wiki.qt.io/Automating_generation_of_qm_files.

In this section, you learned how to deploy your translation files. In the next section, we will summarize our takeaways from this chapter.

Summary

In this chapter, we took a look at the core concepts of internationalization and localization in Qt. We discussed different tools provided by Qt for internationalization. We learned how to use Qt Linguist. We also looked at how to translate a Qt Widgets application into a different language. Then, we learned how to translate dynamically.

In the latter part of the chapter, we discussed translating a Qt Quick application. Afterward, we learned how to switch languages dynamically in a Qt Quick application. Now you can create an application with multiple languages and share it with your clients or friends in a different geographical region.

In *Chapter 12, Performance Considerations*, we will learn about tools and tricks to optimize performance in a Qt application.

12
Performance Considerations

In this chapter, we will give you an overview of performance optimization techniques and how you can apply them in the context of Qt-based application development. Performance is a very important factor in the success of your application. Performance failures can result in business failures, poor customer relationships, a reduction in competitiveness, and revenue loss. Delaying performance optimization can have a huge cost in terms of your reputation and organizational image. Therefore, it is important to do performance tuning.

You will also learn about performance bottlenecks and how to overcome them. We will discuss different profiling tools to diagnose performance problems, focusing specifically on some popular tools. Then, you will learn how to profile and benchmark performance. The chapter also introduces **Qt Modeling Language** (**QML**) Profiler and Flame Graph to find underlying bottlenecks in your Qt Quick application. You will also learn about some best practices that you should follow while developing your Qt application.

We will discuss the following topics:

- Understanding performance optimization
- Optimizing C++ code
- Using concurrency, parallelism, and multithreading

- Profiling a Qt Quick application using QML Profiler and Flame Graph
- Other Qt Creator analysis tools
- Optimizing graphical performance
- Creating benchmarks
- Different analysis tools and optimization strategies
- Performance considerations for Qt Widgets
- Learning best practices of QML coding

By the end of the chapter, you will have learned to write high-performance optimized code for both C++- and QML-based applications.

Technical requirements

The technical requirements for this chapter include minimum versions of Qt 6.0.0 and Qt Creator 4.14.0 installed on the latest desktop platform such as Windows 10, Ubuntu 20.04, or macOS 10.14.

All the code used in this chapter can be downloaded from the following GitHub link: `https://github.com/PacktPublishing/Cross-Platform-Development-with-Qt-6-and-Modern-Cpp/tree/master/Chapter12/QMLPerformanceDemo`.

> **Important note**
> The screenshots used in this chapter are taken on the Windows platform. You will see similar screens based on the underlying platforms in your machine.

Understanding performance optimization

Performance optimization is done to improve an application's performance. You may be wondering why this is necessary. There are many reasons why an application requires performance optimization. When there is a performance problem reported by your users or the **quality assurance** (**QA**) team, the developers may discover something affecting the overall application performance. This may occur due to underlying hardware limitations, poor implementation of code, or scalability challenges.

Optimization is part of the application development process. This can involve optimizing code for performance or optimization for memory usage. Optimization aims to optimize an application's behavior so that it satisfies the product requirements for speed, memory footprint, power usage, and so on. As a result, optimization is almost as crucial as coding functionality in the production phase. Customers may report performance problems as glitches, slow response, and missing functionalities. A faster application executes more efficiently while consuming fewer resources and can handle more tasks in the same amount of time as a slower application. In today's competitive world, faster software means a competitive advantage over rivals. Performance matters a lot on embedded and mobile platforms, with factors such as speed, memory, and power consumption being prevalent.

In a Waterfall process, performance improvement is carried out after application development, during the integration and verification phase. However, in today's Agile world, code performance should be evaluated every couple of sprints for overall application performance. Performance optimization is a continuous process, whereas defect fixing is a one-time task. It is an iterative process in which you will always find something to improve and there will be always scope for improvement in your application. According to the **Theory of Constraints** (**TOC**), there is typically one problem in a complex application that restricts the application from achieving its optimal performance. Such constraints are known as **bottlenecks**. An application's top performance is limited by bottlenecks, hence you should consider performance optimization during your application development life cycle. If ignored, your new product may become a complete disaster and may even ruin your reputation.

Before you jump into optimization, you should define a goal. Then, you should identify the bottleneck or the constraint. After that, think about how you can fix the constraint. You can improve your code and re-evaluate the performance. If it doesn't meet the set goal, you need to repeat the process. However, remember that premature optimization can be the root of all evil. You should implement the primary functionalities first before validating your product and implementing early users' feedback. Remember to make the application run first, then make its functionalities right, and then make it faster.

When you set a performance goal, you need to choose the right technique. There can be multiple goals, such as faster launch time, a smaller application binary, or less **random-access memory** (**RAM**) usage. One goal can impact another goal, so you have to find a balance based on the expected criteria—for example, optimizing the code for performance may impact memory optimization. There may be different ways to improve overall performance; however, you should also follow the organizational coding guidelines and best practices. If you are contributing to an open source project or are a freelance application developer, you should follow standard coding practices to maintain overall code quality.

Some of the important tricks we will be following for performance improvement are listed as follows:

- Using better algorithms and libraries
- Using optimal data structures
- Allocating memory responsibly and optimizing memory
- Avoiding unnecessary copying
- Removing repeated computation
- Increasing concurrency
- Using compiler binary optimization flags

In the following sections, we will discuss opportunities to improve overall application performance in our C++ code.

Optimizing C++ code

In most Qt applications, a significant part of the coding is done in C++, hence you should be aware of C++ optimization tricks. This section is about implementing some of the best practices while writing C++ code. When C++ implementations are written without optimization, they run slowly and consume a lot of resources. Better optimization of your C++ code also offers better control over memory management and copying. There are many opportunities to improve algorithms, ranging from small logical blocks to using **Standard Template Libraries (STLs)**, to writing better data structures and libraries. There are several excellent books and articles on this topic. We will be discussing a few important points for running code faster and using fewer resources.

Some of the important C++ optimization techniques are listed here:

- Focus on algorithms, not on micro-optimization
- Don't construct objects and copy unnecessarily
- Use C++11 features such as move constructor, lambdas, and the `constexpr` functions
- Choose static linking and position-dependent code
- Prefer 64-bit code and 32-bit data
- Minimize array writes and prefer array indexing to pointers
- Prefer regular memory access patterns

- Reduce control flow

- Avoid data dependencies

- Use optimal algorithms and data structures

- Use caching

- Use precomputed tables to avoid repeated computation

- Prefer buffering and batching

Since this book requires previous knowledge of C++, we expect that you will be aware of these best practices. As a C++ programmer, always stay updated with the latest C++ standards such as C++17 and C++20. These will help you in writing efficient code with great features. We won't be discussing these in detail in this section, but will leave this for your self-exploration.

You can read more about C++ core guidelines at the following link: `https://isocpp.github.io/CppCoreGuidelines/CppCoreGuidelines`.

You can learn more about optimizing C++ code at the following link: `https://www.agner.org/optimize/`.

Go through the listed approaches to improve your C++ code. Next, we will discuss how to improve application performance with concurrency and multithreading in the next section.

Using concurrency, parallelism, and multithreading

Since you are already a C++ developer, you might be aware of these terms, which may be used interchangeably. However, there are differences in these terms. Let's revisit these terms here:

- **Concurrency** is the execution of multiple programs at the same time (concurrent).

- **Parallelism** is the simultaneous running of a portion of your program in parallel, utilizing the multiple cores in a multi-core processor.

- **Multithreading** is the capability of a **central processing unit** (**CPU**) to run multiple threads for the same program, concurrently supported by the operating system.

For example, you may launch multiple instances of a **Portable Document Format (PDF)** reader and Qt Creator. Qt Creator can run multiple tools by itself. Your system Task Manager can show you all the processes running simultaneously. This is known as concurrency. It is also commonly known as multitasking.

But if you use parallel computing techniques to process your data, then this is called parallelism. Complex applications with huge data processing requirements use this technique. Note that parallel computing on a single-core processor is an illusion.

A thread is the smallest executable unit of a process. There can be several threads in a process, but there is only one main thread. Multithreading is concurrency within the same process. Only one core is used by a conventional single-threaded application. A program with multiple threads can be distributed to multiple cores, allowing true concurrency. As a result, a multithreaded application provides better performance on multi-core hardware.

Let's discuss a few important classes in Qt that provide concurrency and multithreading, as follows:

- `QThread` is used to manage one thread of control within a program.
- `QThreadPool` is used to manage and recycle individual `QThread` objects to help reduce thread creation costs in a multithreaded application.
- `QRunnable` is an interface class for representing a task or piece of code that needs to be executed.
- `QtConcurrent` offers high-level **application programming interfaces (APIs)** that help in writing multithreaded programs without using low-level threading primitives.
- `QFuture` permits threads to be synchronized against multiple computational results that will be available at a later point in time.
- `QFutureWatcher` provides information and notifications about a `QFuture` object using signals and slots.
- `QFutureSynchronizer` is a convenience class that simplifies the synchronization of one or more `QFuture` objects.

Threads are primarily used in two scenarios, as follows:

- To make use of multi-core CPUs to speed up processing
- Offload long-running processing or blocking calls to other threads to keep the **graphical user interface (GUI)** thread or other time-critical threads responsive

Let's briefly discuss the most basic concurrency concept known as a **thread**. The QThread class offers a thread abstraction in Qt with convenience methods. You can start a new custom thread by subclassing the QThread class, as follows:

```
class CustomThread : public QThread
{
    public:
    void run(){…}
};
```

You can create a new instance of this class and invoke its start() function. This will create a new thread and then call the run() function in the context of this new thread. Another approach is to directly create a QThread object and invoke the start() function, which will start an event loop. In comparison to a conventional C++ thread class, QThread supports thread interruption, which isn't supported in C++11 and later. You may wonder why we can't just use the C++ standard thread class. This is because you can use the signals and slots mechanism with QThread in a multithread-safe way.

You can also use the multithreading mechanism in QML using WorkerScript. JavaScript code can execute in parallel with the GUI thread using the WorkerScript QML type. To enable the use of threads in a Qt Quick application, import the module as follows:

```
import QtQml.WorkerScript
```

One JavaScript can be attached to each WorkerScript object. The script will run in a different thread and QML context when WorkerScript.sendMessage() is called. When the script is finished, it can send a response to the GUI thread, invoking the WorkerScript.onMessage() signal handler. You can exchange data between threads by using signals and signal handlers. Let's have a look at a simple WorkerScript usage, as follows:

```
WorkerScript {
    id: messagingThread
    source: "messaging.mjs"
    onMessage: (messageObject) => textElement.text =
            messageObject.reply
}
```

The preceding code snippet uses a JavaScript file, `messaging.mjs`, which performs the operations in a new thread. Let's look at the sample script, as follows:

```
WorkerScript.onMessage = function(message) {
    //Perform complex operations here
    WorkerScript.sendMessage({ 'reply': 'Message '+
                               message})
}
```

You can send a message from the click of a button or based on some user action. It will invoke the `sendMessage(jsobject message)` method, where your complex messaging operations will take place. You can read more about different threading mechanisms and use cases at the following link: `https://doc.qt.io/qt-6/threads-technologies.html`.

Since this book is written for experienced C++ developers, it is expected that you will be familiar with terms such as `mutex`, `semaphore`, `read-write lock`, and so on. Qt provides convenience classes to use these mechanisms while implementing a multithreading application. We won't deep dive into these Qt classes with examples. You can learn more about the use of `QMutex`, `QSemaPhore`, `QReadWriteLock`, and `QWaitCondition` at the following link: `https://doc.qt.io/qt-6/threads-synchronizing.html`.

In this section, we learned how concurrency mechanisms can be used to improve overall application performance. Don't implement it unnecessarily for simple tasks as this may result in degraded performance. In the next section, we will discuss the use of the QML Profiler tool for profiling a Qt Quick application.

Profiling a Qt Quick application using QML Profiler and Flame Graph

QML in Qt 6 takes advantage of **graphics processing units** (**GPUs**) and uses hardware acceleration for rendering. This feature makes QML superior to Qt Widgets in terms of performance. However, there can be bottlenecks in your QML code that may impact overall application performance. In this section, we will focus on using the built-in tool to find these bottlenecks. Qt Creator provides seamless integration with multiple tools. The most important tool is **QML Profiler**. It is provided by Qt and works on all Qt-supported platforms. Other than QML Profiler, Qt Creator also provides third-party tools such as **Valgrind**, **Heob**, and **Performance Analyzer**. You can enable new plugins or remove some plugins from **About Plugins…**, available under the **Help** menu.

Let's discuss QML Profiler, which you will be using most of the time to find the bottlenecks in your QML code. The goal of QML Profiler is to help you identify bottlenecks by providing you with details such as the time taken by a code block to do a certain operation, after which you can decide to reimplement the code with suitable GUI elements or better data structures or algorithms.

Follow these steps to start profiling and optimizing your Qt Quick application:

1. Open an existing Qt Quick project or create a new Qt Quick application using Qt Creator's **New Project** creation wizard.

2. Once the project is created, add some code to it. Then, select **QML Profiler** under the **Analyze** menu to run the QML Profiler tool. The **Analyze** context menu can differ from platform to platform based on installed plugins. The following screenshot shows the **QML Profiler** option in the Windows platform. In Linux, you may see a few more options, such as **Valgrind Memory Analyzer**, **Valgrind Memory Analyzer with GDB**, and **Valgrind Function Profiler**:

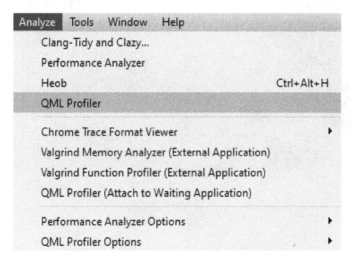

Figure 12.1 – QML Profiler option in Qt Creator integrated development environment (IDE)

3. When you hit the **QML Profiler** option, your Qt Quick application will run by QML Profiler. You will see the **QML Profiler** window appear below the code editor. You may also see the following message:

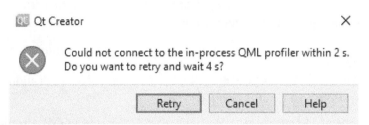

Figure 12.2 – QML Profiler retry message

4. If you get this popup, just hit **Retry**. You will notice that the profiling will begin and you will also notice the output screen. In the sample application, we are creating new rectangles on a mouse click, as illustrated in the following screenshot:

Figure 12.3 – Output of sample Qt Quick application

5. On the **user interface** (**UI**), perform some user interactions—such as click a button—to do a certain operation. Then, click the **Stop** button located on the title bar of the profiler window. You will also see two more buttons on both sides of the **Stop** button. If you hover your mouse over them, you will see their functionalities, such as **Start QML Profiler analysis** and **Disable Profiling**.

An overview of the **QML Profiler** window is shown in the following screenshot:

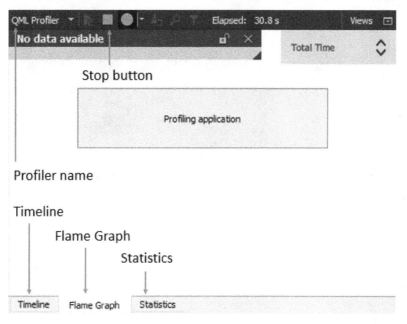

Figure 12.4 – QML Profiler window showing Stop button and tabbed views

6. Once you stop the profiler, you will see the **QML Profiler** window is updated with some views. You will notice that there are three tabs under the profiler window—namely **Timeline**, **Flame Graph**, and **Statistics**.

7.   Let's look at the first tab on QML Profiler—click on the **Timeline** tab. The following
     screenshot shows a sample view of the output:

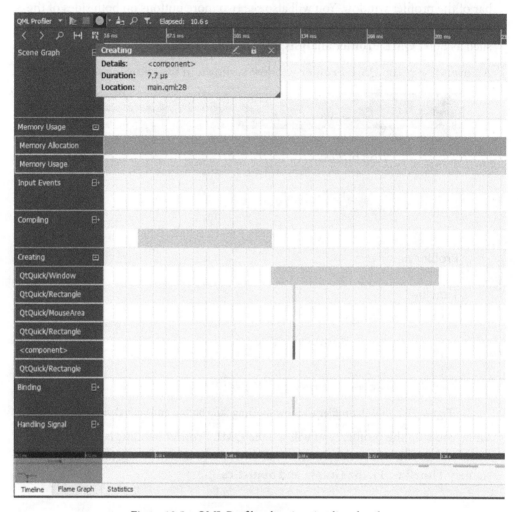

Figure 12.5 – QML Profiler showing timeline details

You will notice that there are six different sections under the timeline display: **Scene
Graph**, **Memory Usage**, **Compiling**, **Creating**, **Binding**, and **JavaScript**. These
sections give us an overview of the different stages of application processing such
as compilation, component creation, and logical execution.

8. You can find colorful bars on the timeline. You can use the mouse wheel to zoom in and zoom out on specific timeline sections. You can also move the timeline by pressing the left mouse button at the bottom region of the timeline and move in either direction to locate an area of interest.

The different sections of the **Timeline** tab are illustrated in the following screenshot:

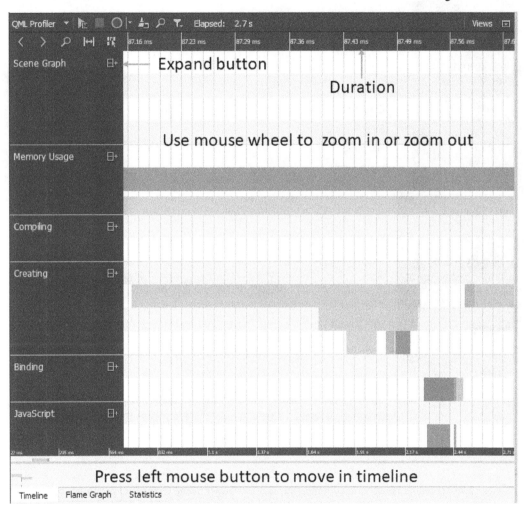

Figure 12.6 – Timeline tab showing different sections

9.   You can click on the **Expand** button to see further details under each section, as illustrated in the following screenshot:

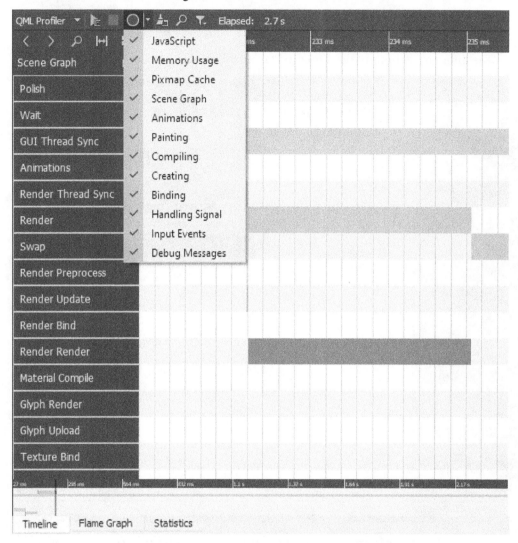

Figure 12.7 – Timeline tab showing different subsections under Scene Graph and profiling options

10. If you click on one of the bars under the **Creating** section, you can find component details such as the QtQuick/Rectangle type, total duration taken for creating an object, and the location of code displayed on a pop-up window, over the **QML Profiler** window. You can use the yellow arrows in the top-left corner to jump to previous or next events. This section is illustrated in the following screenshot:

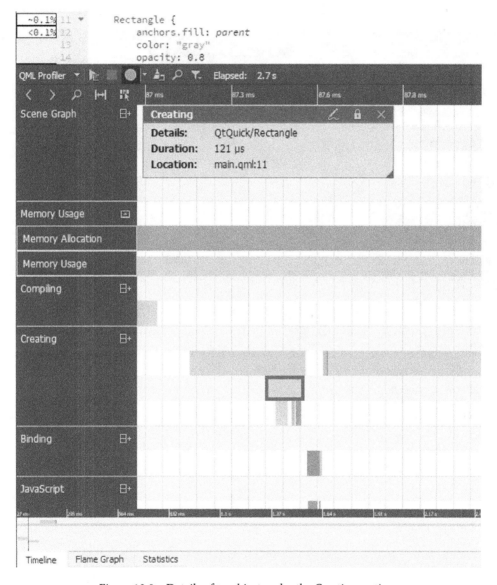

Figure 12.8 – Details of an object under the Creating section

11. You can switch between different tabs at the bottom of the **QML Profiler** window. Once you have explored the **Timeline** tab, let's open up the **Flame Graph** tab. Under this tab, you will find a visualization of the **Total Time**, **Memory**, and **Allocations** of your application as a percentage. You can switch between these views by clicking on the dropdown located in the top-right corner of the **QML Profiler** window, as shown in the following screenshot:

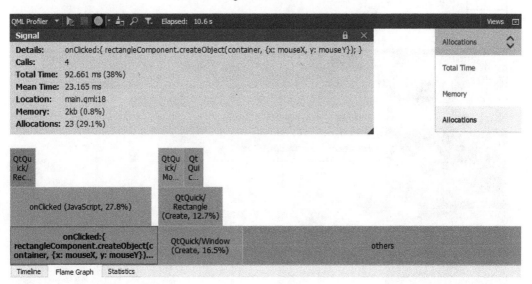

Figure 12.9 – Flame Graph showing Allocations view

12. The **Flame Graph** view provides a more compact statistics summary. The horizontal bars depict one aspect of the samples gathered for a certain function in comparison to the same aspect of all samples combined. The nesting indicates a call tree that shows, for example, which functions call the other function.

13. As seen in the following screenshot, you can also see the percentage value displayed on the left side of the code editor. Based on which component is consuming more time, you can tweak your code:

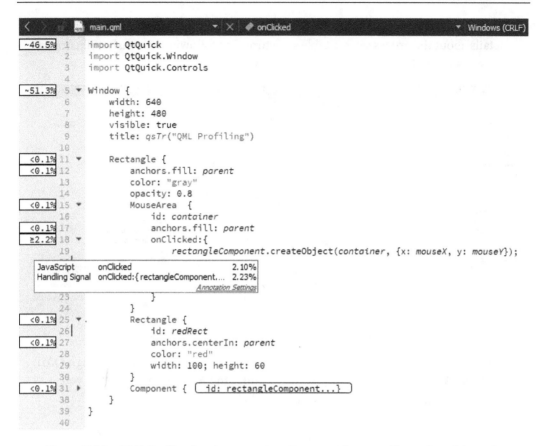

Figure 12.10 – QML Profiler showing percentage time spent for a specific portion of the code

14. Since data collection takes time, you may notice a little lag before the data is displayed. When you click the **Enable Profiling** button, data is transferred to QML Profiler, therefore don't terminate the application immediately.

15. To disable the automatic start of data collection when an application is launched, select the **Disable Profiling** button. When you toggle the button, data collection will start again.

16. Let's move to the next tab: the **QML Profiler** window. This tab reveals statistical details about the processes in a table structure. The following screenshot illustrates the statistics of the code execution for our sample code:

Location	Type	Time in Percent	Total Time	Self Time in Percent	Self Time	Calls	Mean Time	Details
<program>		100 %	244 ms	0.00 %	0 ns	1	244 ms	Main program
main.qml:18	Handling Signal	37.98 %	92.7 ms	36.87 %	90 ms	4	23.2 ms	onClicked:(rectangleComponent.createObject...
main.qml:5	Creating	34.21 %	83.5 ms	34.15 %	83.3 ms	2	41.7 ms	QtQuick/Window
main.qml:0	Compiling	27.79 %	67.8 ms	27.79 %	67.8 ms	1	67.8 ms	main.qml
main.qml:18	JavaScript	1.11 %	2.71 ms	1.07 %	2.61 ms	4	677 µs	onClicked
main.qml:11	Creating	0.05 %	130 µs	0.03 %	70.7 µs	2	65 µs	QtQuick/Rectangle
main.qml:30	Creating	0.04 %	95.6 µs	0.04 %	95.6 µs	8	11.9 µs	QtQuick/Rectangle
main.qml:15	Creating	0.02 %	43.3 µs	0.02 %	43.3 µs	2	21.6 µs	QtQuick/MouseArea
main.qml:12	Binding	0.02 %	41.1 µs	0.00 %	6.2 µs	1	41.1 µs	anchors.fill: parent
main.qml:12	JavaScript	0.01 %	34.9 µs	0.01 %	34.9 µs	1	34.9 µs	expression for fill
main.qml:22	Creating	0.01 %	23.4 µs	0.01 %	23.4 µs	2	11.7 µs	QtQuick/Rectangle
main.qml:28	Creating	0.00 %	7.7 µs	0.00 %	7.7 µs	1	7.7 µs	<component>
<bytecode>	Binding	0.00 %	6.4 µs	0.00 %	6.4 µs	1	6.4 µs	Source code not available
main.qml:24	Binding	0.00 %	1.4 µs	0.00 %	600 ns	1	1.4 µs	anchors.centerIn: parent
main.qml:17	Binding	0.00 %	1.3 µs	0.00 %	600 ns	1	1.3 µs	anchors.fill: parent
main.qml:24	JavaScript	0.00 %	800 ns	0.00 %	800 ns	1	800 ns	expression for centerIn

Caller	Type	Total Time	Calls	Caller Description	Callee	Type	Total Time	Calls	Callee Description
<program>		83.5 ms	1	Main Program	main.qml:11	Creating	130 µs	1	QtQuick/Recta...

Timeline   Flame Graph   Statistics

Figure 12.11 – QML Profiler showing statistics of code execution

17. You can also attach QML Profiler to an externally started application through **QML Profiler (Attach to Waiting Application)** under the **Analyze** menu. Once you select the option, you will see the following dialog:

Figure 12.12 – QML Profiler showing remote execution option

18. To save all of the data collected, right-click on any QML Profiler view and select **Save QML Trace** in the context menu. You can select **Load QML Trace** to see the saved data. You can also send the saved data to others for review or load data that they have saved.

In this section, we discussed different options available in QML Profiler. By using this tool, you can easily find code that is causing performance issues. More details are available at this link: `https://doc.qt.io/qtcreator/creator-qml-performance-monitor.html`.

In the next section, we will discuss further how to use other analytical tools to optimize your Qt code.

Other Qt Creator analysis tools

In the earlier section, we discussed QML Profiler, but you may need to analyze your C++ and Qt Widgets code. Qt Creator provides integration with some of the famous analysis tools to help you analyze your Qt application. Some of the tools that come with Qt Creator are listed here:

- **Heob**
- **Performance Analyzer**
- **Valgrind**
- **Clang Tools: Clang-Tidy and Clazy**
- **Cppcheck**
- **Chrome Trace Format (CTF)** visualizer

Let's briefly discuss these tools and become familiar with them before getting into their documentation.

To use Heob, you first need to download and install it. Buffer overruns and memory leaks can be easily detected with Heob. It works by overriding the caller process's heap functions. An access violation is raised when a buffer overrun occurs, and stack traces of the offending code and buffer allocation are noted. You will find the stack traces when the application exits normally. It doesn't require any recompilation or relinking of the target application.

You can read about its usage on the official documentation link at `https://doc.qt.io/qtcreator/creator-heob.html`.

You can download the binary from `SourceForge.net` or build it from the source code. The source code of Heob can be found at the following link: `https://github.com/ssbssa/heob`.

The Linux Performance Analyzer tool is integrated with Qt Creator and can be used to analyze an application's CPU and memory utilization on Linux desktop or Linux-based embedded systems. The `perf` tool takes periodic snapshots of an application's call tree and visualizes them in a timeline view or as a flame graph, using the utility included with the Linux kernel. You can launch it on your Linux machine from the **Performance Analyzer** option under the **Analyze** menu, as illustrated in the following screenshot:

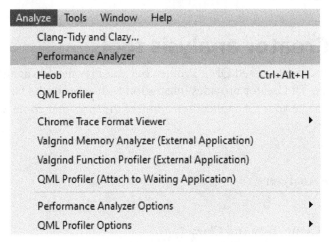

Figure 12.13 – Qt Creator showing Performance Analyzer option

Please note that the **Performance Analyzer** doesn't work on the Windows platform. Even on Linux distributions, if it can't locate the `perf` utility, you will get an equivalent warning dialog, as shown in the next screenshot:

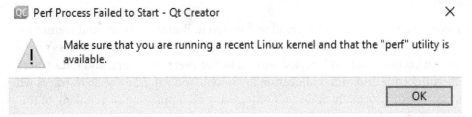

Figure 12.14 – Qt Creator showing Performance Analyzer warning dialog

Use the following command to install the `perf` tool on your Ubuntu machine:

```
$sudo apt install linux-tools-common
```

If you are using a different Linux distribution, you can use the corresponding command. `perf` may fail for the specific Linux kernel, with a warning about the kernel version. In that case, type the following command with the appropriate kernel version:

```
$sudo apt install linux-tools-5.8.0-53-generic
```

Once the `perf` setup is done, you can see the predefined events in the command prompt with the following command:

```
$perf list
```

Next, launch Qt Creator and open a Qt project. Select **Performance Analyzer** from the **Analyze** menu. **Performance Analyzer** will start collecting data as soon as you start examining an application, and the **Recorded** field will show the duration details. Since the data is processed through the `perf` tool and an additional assistance program is included with Qt Creator, it may appear in Qt Creator several seconds after it was created. The **Processing delay** field contains an estimate for this delay. The data collection continues until you click the **Stop collecting profile data** button or close the application.

You can also load `perf.data` and analyze an application from **Performance Analyzer Options** under the **Analyze** menu, as shown here:

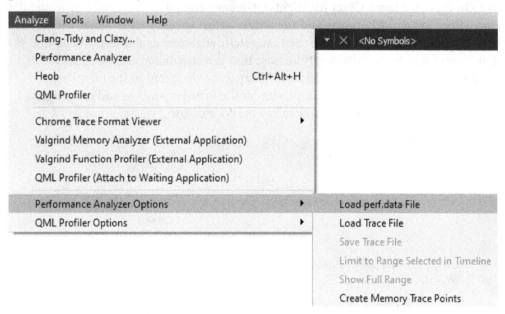

Figure 12.15 – Context menu showing Performance Analyzer options

You can read more about usage of Performance Analyzer at the following link: `https://doc.qt.io/qtcreator/creator-cpu-usage-analyzer.html`.

On macOS, there is an equivalent tool called **Instructions**; however, this is not integrated with Qt Creator. You can launch it separately and look at the **Time Profiler** section.

On Linux and macOS, **Valgrind** is the tool of choice for debugging a wide range of problems. Individual techniques, such as profiling and memory checking, are used for specialized analysis. The **Analyze** menu in Qt Creator combines Valgrind and allows memory testing and profiling from within the IDE. To use Valgrind, it must be installed. It isn't available on Windows. However, since memory problems aren't often platform-specific, you can do analysis on Linux or macOS. **KCachegrind** is the visualizer for Valgrind profiling results. When you run Valgrind, you will notice the profiler window open with memcheck. You can change this to callgrind from the profiler drop-down option.

You can learn more about Valgrind at the following link: `https://doc.qt.io/qtcreator/creator-valgrind-overview.html`.

The next tool available in Qt Creator is **Clang-Tidy** and **Clazy...**. These tools can be used to locate issues in your C++ code through static analysis. **Clang-Tidy** provides diagnostics and fixes for common programming errors such as style violations or interface misuse. On the other hand, **Clazy** highlights Qt-related compiler errors, such as wasteful memory allocation and API usage, and suggests refactoring activities to remedy some of the problems. Clang-Tidy includes the Clang static analyzer capabilities. You do not need to set up Clang tools individually because they are distributed and integrated with Qt Creator. When you run **Clang-Tidy and Clazy...**, as illustrated in the following screenshot, you will see the analysis details under the **Profiler** window and the progress under the **Application Output** window below the code editor:

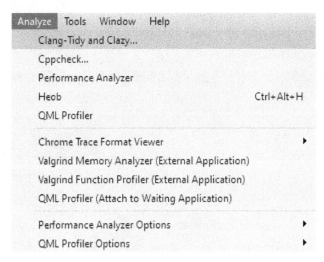

Figure 12.16 – Context menu showing the Clang-Tidy and Clazy... option

Let's run the tool on an existing Qt example. In the application window, you will see the analysis running, and in the profiler window, you will see the results.

You can explore the documentation further at the following link: `https://doc. qt.io/qtcreator/creator-clang-tools.html`.

Qt Creator also includes another tool called **cppcheck**. This tool has experimental integration with Qt Creator. You can enable it from **About Plugins…**, available under the **Help** menu. You can use this to detect undefined behavior and dangerous coding constructs. The tool provides options to check warnings, style, performance, portability, and information.

The last analysis tool integrated with Qt Creator is the **CTF visualizer**. You can use this along with QML Profiler. Tracing information might provide you further insight into the data that QML Profiler collects. You can find why a simple binding is taking so long, such as being possibly impacted by the C++ code or by slow disk operation. Full stack tracing may be used to trace from the top-level QML or JavaScript down to C++ and all the way down to the kernel area. This allows you to assess an application's performance and determine if poor performance is caused by the CPU or other programs on the same system. Tracing provides insight into what the system is doing and why an application is behaving in an undesired way. To see Chrome trace events, utilize the CTF visualizer.

You can learn more about the CTF visualizer at the following link: `https://doc. qt.io/qtcreator/creator-ctf-visualizer.html`.

In this section, we have discussed different analysis tools available in Qt Creator. In the next section, we will discuss further how to optimize and locate graphical performance issues.

Optimizing graphical performance

We discussed graphics and animation in *Chapter 8, Graphics and Animations*. In this section, we will explore factors that impact performance in graphics and animation. Graphics performance is essential in any application. If your application is poorly implemented, then users may see flickering in the UI or the UI may not update as expected. As a developer, you must make every effort to ensure that the rendering engine maintains a 60 **frames-per-second** (**FPS**) refresh rate. There are only 16 **milliseconds** (**ms**) between each frame in which processing should be done at 60 FPS, which includes the processing necessary to upload the draw primitives to the graphics hardware.

To avoid any glitch in graphical performance, you should use asynchronous, event-driven programming wherever possible. If your application has huge data processing requirements and complex calculations, then use worker threads to do the processing. You should never manually spin an event loop. Don't spend more than a few ms per frame in blocking functions. If you don't follow these points, the users will see the GUI flickering or freezing, resulting in a bad **user experience (UX)**. When it comes to generating graphics and animations on the UI the QML engine is very efficient and powerful. However, there are a few tricks you can use to make things even go faster. Instead of writing your own, utilize Qt 6's built-in capabilities.

While drawing graphics, you should choose opaque primitives if possible. Opaque primitives are faster to render by the renderer and to draw on the GPU. Hence, between **Portable Network Graphics (PNG)** and **Joint Photographic Experts Group (JPEG)** files, rendering JPEG formats is faster. You should be using `QImage::Format_RGB32` when passing photos to a `QQuickImageProvider`. Please note that overlapping compound items cannot be batched. Avoid clipping if possible as it breaks batching. Instead of clipping an image, use `QQuickImageProvider` to generate a cropped image. Applications that require a monochromatic background should use `QQuickWindow::setColor()` rather than a top-level `Rectangle` element. `QQuickWindow::setColor()` invokes `glClear()`, which is faster.

While using `Image`, make use of the `sourceSize` property. The `sourceSize` property enables Qt to downsize the image before loading it into memory, ensuring that huge images consume no more memory than is required. When the `smooth` attribute is set to `true`, Qt filters the image to make it look smoother when it is scaled or changed from its original size. If the image is rendered at the same size as its `sourceSize` property, this makes no difference. On some older hardware, this property will influence the performance of your application. The `antialiasing` property directs Qt to smooth down aliasing artifacts around the edges of the image. This property will affect your program's performance.

Better graphical performance can be achieved by effective batching. The renderer can provide statistics on how well the batching runs, how many batches are utilized, which batches are kept, which are opaque, and which are not. To enable this, add an environment variable such as `QSG_RENDERER_DEBUG` and set the value to `render`. Unless an image is too huge, a texture atlas is used by the `Image` and `BorderImage` QML types. If you are creating textures using C++, then call `QQuickWindow::createTexture()` and pass `QQuickWindow::TextureCanUseAtlas`. You can use another environment variable, `QSG_ATLAS_OVERLAY`, to colorize the atlas textures, which helps in identifying them easily.

To visualize the various aspects of the scene graph's default renderer, the `QSG_VISUALIZE` environment variable can be set to one of the values. You can do this in Qt Creator by going to the **Projects** tab, expanding the **Build Environment** section, clicking **Add**, then entering the variable name as `QSG_VISUALIZE` and setting the value for that variable, as follows:

- `QSG_VISUALIZE = overdraw`
- `QSG_VISUALIZE = batches`
- `QSG_VISUALIZE = clip`
- `QSG_VISUALIZE = changes`

When `QSG_VISUALIZE` is set to `overdraw`, overdraw is visualized in the renderer. To highlight overdraws, all elements are visualized in **three dimensions (3D)**. To some extent, this mode may also be used to identify geometry outside the viewport. Translucent items are shown with a red tint, whereas opaque items are shown with a green tint. The viewport's bounding box is shown in blue. Don't use `Rectangle` just to draw a white background, as `Window` also has a white background. In this case, using an `Item` property instead of `Rectangle` can improve performance.

Setting `QSG_VISUALIZE` to `batches` causes batches to be visualized in the renderer. Unmerged batches are drawn with a diagonal line pattern, whereas merged batches are drawn with a solid color. A small number of distinct colors indicates effective batching. Unmerged batches are undesirable if they contain a large number of individual nodes.

All QML components that derive from `Item` have a property called `clip`. By default, the `clip` value is set to `false`. This property informs the scene graph not to render any child elements that extend beyond the boundaries of their parent. When `QSG_VISUALIZE` is set to `clip`, red spots appear on top of the scene to indicate clipping. Because Qt Quick `Items` do not clip by default, clipping is often not shown. Clipping prevents the ability to batch multiple components together, which impacts graphical performance.

When `QSG_VISUALIZE` is set to `changes`, changes in the renderer are shown. A flashing overlay of random color is used to highlight changes in the scene graph. Modifications to a primitive are shown by a solid color, but changes to an ancestor— such as changes to the matrix or opacity—are shown by a pattern.

Experiment with these environment variables in your Qt Quick application. You can learn more about these rendering flags at the following link: `https://doc.qt.io/qt-6/qtquick-visualcanvas-scenegraph-renderer.html`.

Qt Quick helps in building great applications with a fluid UI and dynamic transitions. However, you should consider some of the factors to avoid performance implications. When you add an animation to a property, all bindings are impacted and re-evaluated, which references the property. To avoid performance issues, you may remove the binding before running the animation and then reassign it after the animation is complete. During the animation, avoid using JavaScript. Script animations should be used with caution because they run in the main thread.

You can use Qt Quick particles to create a nice particle effect. However, its performance depends on underlying hardware capabilities. To render more particles, you will need faster graphics hardware. Your graphics hardware should be capable to draw at or above 60 FPS. You can learn more about optimizing particle performance at the following link: `https://doc.qt.io/qt-6/qtquick-particles-performance.html`.

In this section, we discussed different considerations to optimize graphical performance. In the next section, we will discuss further how to benchmark your application.

Creating benchmarks

We have learned about benchmarking in *Chapter 9*, *Testing and Debugging*. Let's look at some aspects of benchmarking to evaluate performance issues. We've already talked about Qt Test's support for benchmarking, which is a calculation of the average time required by a particular task. The QBENCHMARK macro is used to benchmark a function.

The following code snippet shows benchmarking key clicks on a line edit:

```
void LineEditTest::testClicks()
{

    auto tstLineEdit = ui->lineEdit;
    QBENCHMARK {QTest::keyClicks(tstLineEdit, "Some
                Inputs");}
}
```

You can also benchmark convenience functions provided by Qt. The following code benchmarks the QString::localeAwareCompare() function. Let's look at the sample code here:

```
void TestQStringBenchmark::simpleBenchmark()
{

    QString string1 = QLatin1String("Test string");
    QString string2 = QLatin1String("Test string");
```

```
    QBENCHMARK {string1.localeAwareCompare(string2);}
}
```

You can also run benchmark tests in QML. The Qt benchmark framework will run functions with names that begin with `benchmark_` several times, with an average timing value recorded for the runs. It is similar to the `QBENCHMARK` macro in the C++ version of `QTestLib`. You can prefix the test function name with `benchmark_once_` to get the effect of the `QBENCHMARK_ONCE` macro.

You can also use the **qmlbench** tool provided by Qt Labs. This is a benchmarking tool that evaluates your Qt application as a single stack rather than in isolation, and the benchmarks give a lot of insight into the overall performance of your Qt application. It has several readymade shells that come with built-in benchmarking logic. You can do two different types of benchmarking with qmlbench, such as plain `Benchmark` or `CreationBenchmark`. It also allows you to perform both automated and manual benchmarking. Automated tests can be used for regression testing, whereas manual tests can be done to understand the capabilities of new hardware. It comes with built-in features such as the FPS counter, which is very important for GUI applications. You can find the frame rate by running the following command:

```
>qmlbench --shell frame-count
```

You can also run all the automated tests with a simple command, as follows:

```
>qmlbench benchmarks/auto/
```

To explore more about the tool and look at the examples, please refer to the following link: `https://github.com/qt-labs/qmlbench`.

We have seen benchmarking object creation in Qt Widgets and QML and we also benchmarked a Qt function. You can also analyze without using any macros. You can simply use `QTime` or `QElapsedTimer` to measure the time taken by a portion of a code or a function, as illustrated in the following code snippet:

```
QTime* time = new QTime;
time->start();
int lastElapsedTime = 0;
qDebug()<<"Start:"<<(time->elapsed()-
        lastElapsedTime)<<"msec";
//Do some operation or call a function
qDebug()<<"End:"<<(time->elapsed()-
        lastElapsedTime)<<"msec";
```

In the preceding code snippet, we have used `elapsed()` to measure the time taken for a code segment. The difference is that you can evaluate a few lines inside a function—you don't have to write a separate test project. It's a quick way to find performance issues without evaluating a whole project.

You can also benchmark your Qt Quick 3D application. Here's an article on how to do it: `https://www.qt.io/blog/introducing-qtquick3d-benchmarking-application`.

In this section, we discussed benchmarking techniques. In the next section, we will discuss more profiling tools.

Different analysis tools and optimization strategies

You can optimize your application at multiple levels other than just at a code level. Optimization can also be done at a memory or binary. You can modify your application to make it work more efficiently by using fewer resources. However, there can be a trade-off between memory and performance. Based on your hardware configuration, you can decide a strategy as to whether memory usage or processing time is important. In some embedded platforms with memory limitations, you can allow the processing time to be a little longer to use less memory and keep the application responsive. You can also delegate some part of the optimization task to the compiler.

Let's have a look at different strategies we can use to build, analyze, and deploy faster.

Memory profiling and analysis tools

In this section, we will discuss some additional tools you can use to analyze your application. Note that we won't be discussing these tools in detail. You can visit the respective tool website and learn from their documentation. In addition to the available tools in Qt Creator, you can use the following tools on your Windows machine.

Let's have a look at the list of tools, as follows:

- **AddressSanitizer** (**ASan**) is an address monitoring tool built by Google and part of Sanitizers.
- **AQTime Pro** finds issues and memory leaks with application runtime analysis and performance profiling.

- **Deleaker** is a tool for C++ developers who want to find all possible known leaks in their projects. It can detect memory leaks, **Graphics Device Interface** (**GDI**) leaks, and other leaks.

- **Intel Inspector XE** is a memory and thread debugger from Intel.

- **PurifyPlus** is a runtime analysis tool suite that monitors your program as it runs and reports on key aspects of its behavior.

- **Visual Leak Detector** is a free, robust, open-source memory leak detection system for Visual C++.

- **Very Sleepy** is a CPU profiler based on sampling.

- **Visual Studio Profiler** (**VSTS**) can be used for CPU sampling, instrumentation, and memory allocation.

- **MTuner** utilizes a novel approach to memory profiling and analysis, keeping an entire time-based history of memory operations.

- **Memory Leak Detection Tool** is a high-performance memory leak detection tool.

- **Heob** detects buffer overruns and memory leaks. Integrated into Qt Creator.

- **Process Explorer** can query and visualize several systems and performance counters for each process, and I regularly use it for preliminary investigations.

- **System Explorer** shows all system calls issued by any running processes in a long list and supports filters to select processes we'd like to observe.

- **RAMMap** examines a system's global memory usage, which requires quite a bit of Windows internal knowledge.

- **VMMap** shows detailed information on a single application's memory usage.

- **Coreinfo** gives detailed information about the processor, information you might need when doing low-level optimization work.

- **Bloaty** performs a deep analysis of the binary. It aims to accurately attribute every byte of the binary to a symbol or compile the unit that produced it.

In this section, we briefed you about some of the third-party profiling tools. In the next section, we will discuss how to optimize your binary during linking.

Optimizing during linking

In earlier sections, we discussed how to find bottlenecks and optimize a code segment that is impacting an application's performance. Fortunately, most compilers now include a mechanism that allows you to do such optimizations while maintaining the modularity and cleanliness of your code. This is referred to as **link-time code generation (LTCG)** or **link-time optimization (LTO)**. LTO is the optimization of a program during the linking process. The linker collects all object files and integrates them into a single program. Because the linker can view the entire program, it can do whole-program analysis and optimization. However, the linker generally only sees the program after it has been translated into machine code. Rather than converting each source file to machine code one by one, we postpone the code-generation procedure until the very end—linking time. Code generation at linking time enables not just smart inlining of code but also does optimizations such as devirtualizing functions and better elimination of redundant code. This technique can be used to improve application launch time.

To enable this mechanism in Qt, you have to build from the source code. At the configure step, add `-ltcg` to the command-line options. Compiling all of your source code at once during the compilation stage will provide you all of the optimization benefits of full LTO. You can optimize your application launch time at a toolchain, platform, and application level.

Learn more about these performance tips at the following link: `https://wiki.qt.io/Performance_Tip_Startup_Time`.

You can delegate the optimization task to the compiler at times. When you enable optimization flags, the compiler will try to boost the performance and optimize the code block, at the cost of compilation time and—probably—debugging capability. You can enable compiler-level optimization flags for your desired compilers such as **GNU Compiler Collection (GCC)** or Clang.

Look at GCC optimization options for available C++ compilers at the following link: `http://gcc.gnu.org/onlinedocs/gcc/Optimize-Options.html`.

You can learn about different flags in Clang at the following link: `https://clang.llvm.org/docs/CommandGuide/clang.html`.

In this section, you learned about link-time optimization. In the next section, we will discuss how to build your Qt application faster.

Building a Qt application faster

In a large complex project, the time spent on building a project is increasingly becoming valuable. In general, the longer the build time, the more time you lose every day. If you multiply that by the time for a complete team, you lose a lot of time just waiting for the build to finish. While having to wait hours for each small change to be rebuilt might make you more careful about details and drive you to think about each step in depth, it may also limit a more Agile process or collaboration. In this section, we will provide a short guide for dealing with optimization in C++ using Qt.

Please note the following points you should follow to speed up your build process:

- Use parallel building flags
- Make use of a precompiled header (pch)
- Remove redundant targets from makefile
- Use forward declarations in classes

The most effective way while building a large project is to use a parallel-build approach. A parallel build can be enabled by passing an additional parameter. In Qt Creator, you can enable **Parallel Build** under **Build Settings**. You can find the editable fields starting with the **Make** and **Details** buttons under **Build Steps**. Click on the **Details** button, and in the **Make arguments** field, enter -j8. You can instruct your compiler to build in a parallel way through the following command-line statement:

```
>make -j8
```

The last number depends on your hardware. -j8 instructs to run eight threads in parallel. Based on your machine configuration, you may use -j4.

You can also enable a parallel build for the **Microsoft Visual C++ (MSVC)** compiler by enabling the -MP flag. You can instruct cl to run parallel by adding the following flag in the .pro file:

```
*msvc* {
    QMAKE_CXXFLAGS += -MP
}
```

A precompiled header is an excellent technique to drastically minimize a compiler's load. When a compiler parses a file, it must parse the entire code, along with the standard headers and other third-party sources. pch allows you to define which files are frequently used so that the compiler may precompile them before starting a build and utilize the results while building each .cpp file.

To use a precompiled header file, add the following lines of code to the `.pro` file:

```
PRECOMPILED_HEADER = ../pch/your_precompiled_header.h
CONFIG += precompile_header
```

If you use the `Q_OBJECT` macro, the meta-object compiler generates additional files. Don't use the `Q_OBJECT` macro unnecessarily, unless you require relevant features such as the signals and slots mechanism or translation. When you add the `Q_OBJECT` macro, `moc` will generate a `moc_<ClassName>.cpp` file, which adds to the compilation complexity.

You can include this file at the end of your `.cpp` file, as follows:

```
#include "moc_<ClassName>.cpp"
```

You can also lower the dependencies of each `.cpp` file by using forward declarations for small projects and a forward header in large projects. Forwarding classes will shorten the duration of a partial build during standard work. Most classes can contain forward declarations in the `forwards.h` file. By having such a file, you may drastically minimize the number of includes in header files, usually by including `forwards.h`.

As a result, qmake will notice this and remove this file from the list of targets. This will reduce the load on the compiler.

In this section, you learned how to reduce the application build time. In the next section, we will discuss some of the best practices in the Qt Widgets-based application.

Performance considerations for Qt Widgets

The Qt Widgets module renders widgets utilizing the raster engine, a software renders using CPU rather than GPU. In most cases, it can provide the desired performance. However, the Qt Widgets module is very old and lacks the latest capabilities. Since QML is entirely hardware-accelerated, you should consider adopting it for your application's UI.

If your widgets don't need `mouseTracking`, `tabletTracking`, or similar event capturing, turn it off. Your application will use more CPU time as a result of this tracking. Maintain a smaller style sheet and keep it all in one style sheet instead of applying it to individual widgets. A large style sheet will take longer for Qt to process the information into the rendering system, which may affect the application's performance. Use custom styles instead of a style sheet as this can provide you better performance.

Don't create screens unnecessarily and keep them hidden. Create a screen only when it is required. While using `QStackedWidget`, avoid adding too many pages and populating them with many widgets. It requires Qt to discover them all recursively during the rendering and event handling stages, causing the program to run slowly.

Use asynchronous methods wherever feasible for huge operations, to avoid blocking the main process, and keep your software running smoothly. Multithreading is extremely useful for parallelizing several processes in event loops. However, if not done correctly, such as by repeatedly creating and removing threads or by poorly implemented inter-thread communications, it may result in undesired outcome.

Different C++ containers yield different speeds. Qt's vector container is slightly slower than the one in the STL. Overall, the old C++ array is still the fastest, but it lacks sorting capabilities. Use what is most appropriate for your needs.

In this section, you learned about best practices while using the Qt Widgets module. In the next section, we will discuss best practices in QML.

Learning best practices of QML coding

It is important to follow certain best practices while coding in QML. You should keep the file under a certain line limit and should have consistent indentation and structural attributes, as well as following a standard naming convention.

You can structure your QML object attributes in the following order:

```
Rectangle {
// id of the object
// property declarations
// signal declarations
// javascript functions
// object properties
// child objects
// states
// transitions
}
```

If you are using multiple properties from a group of properties, then use group notation, as shown next:

```
Rectangle {
    anchors {
```

```
            left: parent.left; top: parent.top
            right: parent.right; leftMargin: 20
        }
    }
```

Treating groups of properties as a block can ease confusion and help relate the properties with other properties.

QML and JavaScript do not enforce private properties like C++ does. There is a need to hide these private properties—for example, when the properties are part of the implementation. To effectively gain private properties in a QML item, you can embed inside a `QtObject{...}` to hide the properties. This prevents the properties from being accessed outside of the QML file and JavaScript. To minimize the impact on performance, try to group all private properties into the same `QtObject` scope.

The following code snippet illustrates the use of `QtObject`:

```
Item {
    id: component
    width: 40; height: 40
    QtObject {
        id: privateObject
        property real area: width * height //private
                                           //property
    }
}
```

It takes time for property resolution. While the result of a lookup can sometimes be cached and reused, it is generally preferable to avoid doing extra work if at all feasible. You should try to use the common base just once in a loop.

If any of the properties change, the property binding expression is re-evaluated. If you have a loop where you do some processing but only the result matters, then it is better to create a temporary accumulator then assign it to the property you want to update, rather than incrementally updating the property itself, to prevent triggering re-evaluation of binding expressions.

To prevent a continuous overhead of leaving items that are invisible because they are children of a non-visible active element, they should be initialized lazily and destroyed when no longer in use. An object loaded using a `Loader` element may be released by resetting the `source` or `sourceComponent` property of `Loader`, but other items can be explicitly destroyed. It may be required to keep the item active in some situations, in which case it should be made invisible.

In general, opaque content is much faster to draw than translucent content. The reason for this is that translucent content requires blending, and the renderer may be able to better optimize opaque content. Even if an image has only one translucent pixel, it is viewed as totally transparent. The same may be said for a `BorderImage` element with translucent edges.

Avoid doing long logical calculations in QML. Use C++ for implementing business logic. If you still need to use JavaScript-based implementation for doing some complex operation or processing, then use `WorkerScript`.

The Qt Quick Compiler lets you compile QML source code into a final binary. The application's launch time can be greatly reduced by enabling this. You do not have to deploy the `.qml` files along with the application. You can enable Qt Quick Compiler by adding the following line to your Qt project (`.pro`) file:

```
CONFIG += qtquickcompiler
```

To learn more about Qt Quick best practices, read the documentation at the following link: `https://doc.qt.io/qt-6/qtquick-bestpractices.html`.

You can also explore more about Qt Quick performance in the documentation found at the following link: `https://doc.qt.io/qt-6/qtquick-performance.html`.

In this section, we learned some of the best practices while coding in QML. We will now summarize our learning in this chapter.

Summary

In this chapter, we discussed performance considerations and how to improve your overall application performance. We started with improving C++ code. Then, we explained how concurrency techniques can help in making your application faster. You learned about QML Profiler and other profiling tools. You also understood the importance of using best practices while coding in Qt. Now, you can use these techniques in everyday coding. You don't have to be an extraordinary application developer to do performance optimization. If you follow best practices, design patterns, and write better algorithms, then your application will have fewer defects and fewer customer complaints. It is a continuous process, and you will gradually become better at it.

Congratulations! You have learned the basics of performance optimization. If you are curious to know more, then you can read more books specifically written for performance tuning. Happy coding in Qt. Remember—writing better and high-performant code can reduce the CPU cycle, which in turn reduces the carbon footprint, hence effectively, if you code better, you can save the planet and fight climate change!

`Packt.com`

Subscribe to our online digital library for full access to over 7,000 books and videos, as well as industry leading tools to help you plan your personal development and advance your career. For more information, please visit our website.

Why subscribe?

- Spend less time learning and more time coding with practical eBooks and Videos from over 4,000 industry professionals

- Improve your learning with Skill Plans built especially for you

- Get a free eBook or video every month

- Fully searchable for easy access to vital information

- Copy and paste, print, and bookmark content

Did you know that Packt offers eBook versions of every book published, with PDF and ePub files available? You can upgrade to the eBook version at `packt.com` and as a print book customer, you are entitled to a discount on the eBook copy. Get in touch with us at `customercare@packtpub.com` for more details.

At `www.packt.com`, you can also read a collection of free technical articles, sign up for a range of free newsletters, and receive exclusive discounts and offers on Packt books and eBooks.

Other Books You May Enjoy

If you enjoyed this book, you may be interest ed in these other books by Packt:

Hands-On Embedded Programming with QT

John Werner

ISBN: 978-1-78995-206-3

- Understand how to develop Qt applications using Qt Creator on Linux
- Explore various Qt GUI technologies to build resourceful and interactive applications
- Understand Qt's threading model to maintain a responsive UI
- Get to grips with remote target load and debug using Qt Creator
- Become adept at writing IoT code using Qt
- Learn a variety of software best practices to ensure that your code is efficient

Hands-On High Performance Programming with Qt 5

Marek Krajewski

ISBN: 978-1-78953-124-4

- Understand classic performance best practices

- Get to grips with modern hardware architecture and its performance impact

- Implement tools and procedures used in performance optimization

- Grasp Qt-specific work techniques for graphical user interface (GUI) and platform programming

- Make Transmission Control Protocol (TCP) and Hypertext Transfer Protocol (HTTP) performant and use the relevant Qt classes

- Discover the improvements Qt 5.9 (and the upcoming versions) holds in store

- Explore Qt's graphic engine architecture, strengths, and weaknesses

Packt is searching for authors like you

If you're interested in becoming an author for Packt, please visit authors.
packtpub.com and apply today. We have worked with thousands of developers and
tech professionals, just like you, to help them share their insight with the global tech
community. You can make a general application, apply for a specific hot topic that we
are recruiting an author for, or submit your own idea.

Leave a review - let other readers know what you think

Please share your thoughts on this book with others by leaving a review on the site that
you bought it from. If you purchased the book from Amazon, please leave us an honest
review on this book's Amazon page. This is vital so that other potential readers can see
and use your unbiased opinion to make purchasing decisions, we can understand what
our customers think about our products, and our authors can see your feedback on the
title that they have worked with Packt to create. It will only take a few minutes of your
time, but is valuable to other potential customers, our authors, and Packt. Thank you!

Index

www.ingramcontent.com/pod-product-compliance
Lightning Source LLC
Chambersburg PA
CBHW081459050326
40690CB00015B/2860